IIASA PROCEEDINGS SERIES

Volume 17

Mathematical Models for Planning and Controlling Air Quality

IIASA PROCEEDINGS SERIES

1 CARBON DIOXIDE, CLIMATE AND SOCIETY
Proceedings of an IIASA Workshop Cosponsored by WMO, UNEP, and SCOPE,
February 21–24, 1978
Jill Williams, Editor

2 SARUM AND MRI: DESCRIPTION AND COMPARISON OF A WORLD MODEL AND A NATIONAL MODEL
Proceedings of the Fourth IIASA Symposium on Global Modeling,
September 20–23, 1976
Gerhart Bruckmann, Editor

3 NONSMOOTH OPTIMIZATION
Proceedings of an IIASA Workshop,
March 28–April 8, 1977
Claude Lemarechal and Robert Mifflin, Editors

4 PEST MANAGEMENT
Proceedings of an International Conference,
October 25–29, 1976
G.A. Norton and C.S. Holling, Editors

5 METHODS AND MODELS FOR ASSESSING ENERGY RESOURCES
First IIASA Conference on Energy Resources,
May 20–21, 1975
Michel Grenon, Editor

6 FUTURE COAL SUPPLY FOR THE WORLD ENERGY BALANCE
Third IIASA Conference on Energy Resources,
November 28–December 2, 1977
Michel Grenon, Editor

7 THE SHINKANSEN HIGH-SPEED RAIL NETWORK OF JAPAN
Proceedings of an IIASA Conference,
June 27–30, 1977
A. Straszak and R. Tuch, Editors

8 REAL-TIME FORECASTING/CONTROL OF WATER RESOURCE SYSTEMS
Selected Papers from an IIASA Workshop,
October 18–20, 1976
Eric F. Wood, Editor, with the Assistance of Andràs Szöllösi-Nagy

9 INPUT–OUTPUT APPROACHES IN GLOBAL MODELING
Proceedings of the Fifth IIASA Symposium on Global Modeling,
September 26–29, 1977
Gerhart Bruckmann, Editor

10 CLIMATIC CONSTRAINTS AND HUMAN ACTIVITIES
Selected Papers from an IIASA Task Force Meeting,
February 4–6, 1980
Jesse Ausubel and Asit K. Biswas, Editors

11 DECISION SUPPORT SYSTEMS: ISSUES AND CHALLENGES
Proceedings of an International Task Force Meeting,
June 23–25, 1980
Göran Fick and Ralph H. Sprague, Jr., Editors

12 MODELING OF LARGE-SCALE ENERGY SYSTEMS
Proceedings of the IIASA/IFAC Symposium on Modeling of Large-Scale Energy Systems,
February 25–29, 1980
W. Häfele, Editor, and L.K. Kirchmayer, Associate Editor

13 LOGISTICS AND BENEFITS OF USING MATHEMATICAL MODELS OF HYDROLOGIC AND WATER RESOURCE SYSTEMS
Selected Papers from an International Symposium,
October 24–26, 1978
A.J. Askew, F. Greco, and J. Kindler, Editors

14 PLANNING FOR RARE EVENTS: NUCLEAR ACCIDENT PREPAREDNESS AND MANAGEMENT
Proceedings of an International Workshop,
January 28–31, 1980
John W. Lathrop, Editor

15 SCALE IN PRODUCTION SYSTEMS
Based on an IIASA Workshop,
June 26–29, 1979
John A. Buzacott, Mark F. Cantley, Vladimir N. Glagolev, and Rolfe C. Tomlinson, Editors

16 MANAGING TECHNOLOGICAL ACCIDENTS: TWO BLOWOUTS IN THE NORTH SEA
Proceedings of an IIASA Workshop on Blowout Management,
April 1978
David W. Fischer, Editor

17 MATHEMATICAL MODELS FOR PLANNING AND CONTROLLING AIR QUALITY
Proceedings of an IIASA Workshop,
October 1979
Giogio Fronza and Piero Melli, Editors

MATHEMATICAL MODELS FOR PLANNING AND CONTROLLING AIR QUALITY

Proceedings of an October 1979 IIASA Workshop

GIORGIO FRONZA and PIERO MELLI
Editors

PERGAMON PRESS

OXFORD · NEW YORK · TORONTO · SYDNEY · PARIS · FRANKFURT

U.K. — Pergamon Press Ltd., Headington Hill Hall, Oxford OX3 0BW, England

U.S.A. — Pergamon Press Inc., Maxwell House, Fairview Park, Elmsford, New York 10523, U.S.A.

CANADA — Pergamon Press Canada Ltd., Suite 104, 150 Consumers Rd., Willowdale, Ontario M2J 1P9, Canada

AUSTRALIA — Pergamon Press (Aust.) Pty. Ltd., P.O. Box 544, Potts Point, N.S.W. 2011, Australia

FRANCE — Pergamon Press SARL, 24 rue des Ecoles, 75240 Paris, Cedex 05, France

FEDERAL REPUBLIC OF GERMANY — Pergamon Press GmbH, 6242 Kronberg-Taunus, Hammerweg 6, Federal Republic of Germany

First edition 1982

British Library Cataloguing in Publication Data

Mathematical methods for planning & controlling air quality.
— (IIASA proceedings series; v.17)
1. Air — Pollution — Congresses
2. Mathematical models
I. Fronza, Giorgio II. Melli, Piero III. Series
628.5'3 TD883

ISBN 0-08-029950-4

Printed in Great Britain by A. Wheaton & Co. Ltd., Exeter

PREFACE

Air-quality management problems are characterized by a number of difficulties:

First, it is not easy for the decision maker to obtain a satisfactory picture of the physical phenomenon (dispersion of pollutants in the atmosphere) and, in particular, of its most conspicuous aspect (occurrence of high pollutant accumulations and subsequently of dangerous ground-level concentrations). This is due to the influence of a large number of varying meteorological factors, to the existence of widespread emission sources of different types at different heights, to the occurrence of complex chemical reactions between pollutants (or between pollutants and compounds naturally present in the air), and so on.

Second, a large number of individuals are involved in air pollution: a large number contribute to it and a large number suffer from it. Therefore, it is not easy to assess the costs and benefits (or the social values) of different air-quality strategies. Moreover, evaluating the benefits of an abatement strategy requires an assessment of the role of pollution in damaging health, vegetation, and other elements and an economic evaluation of such damage, which are both usually very difficult tasks.

Third, the technology for abating pollutant emissions is not yet well established. There are many drawbacks, in terms of either cost (both installation and maintenance) or technical efficiency (low performance, production of secondary pollutants, and so on).

The existence of these difficulties is confirmed by the history of air-quality legislation; for example, three major amendments to the US Clean Air Act were passed during a period of only ten years.

A large number of mathematical models concerning air pollution have been produced. The large majority of these are merely descriptive: they supply a quantitative explanation of pollutant dispersion in the atmosphere. However, this unfortunately does not mean that clear-cut conclusions about the significance and range of application of descriptive models have been reached.

On the other hand, formal management models based on mathematical programs and looking for "optimal" decisions are relatively few. The most significant ones can be found in environmental economics literature, although often under the general category of "externality management", and not under the specific heading of air-quality planning.

Our general impression is that, on the one hand, modeling has little impact on actual decision making (for instance, few models are considered by US state implementation plans), and, on the other, there is little concern on the part of descriptive modelers about the potential use of models by decision makers.

This book contains contributed papers from the IIASA Workshop on Mathematical Models for Planning and Controlling Air Quality that took place in October 1979.

The Workshop had two goals – which this book shares. The first goal was to contribute to bridging the gap between air-quality modeling and management. In particular the Workshop examined three questions:

Question I	What are the goals actually pursued by decision makers?
Question II	How can such goals be pursued with the aid of mathematical models; i.e., what is the potential role of models in air-quality management?
Question III	What is the actual impact of models in real decision making?

With respect to Question I, the book offers the viewpoint of decision makers from the Federal Republic of Germany, one of the countries that presently has relatively settled air-quality legislation.

Question II is discussed through descriptions of relevant modeling work, covering the full spectrum of modeling research in the air-pollution area, including its economic and policy-making aspects.

The answers to Question III (see the contributions by Miller and by Dennis) are not encouraging, although not entirely pessimistic.

The second goal of the Workshop was to consider an unusual air-quality control strategy: namely, real-time emission control. Such a strategy is an alternative or complement to standard air-quality planning approaches. Though conceptually simple, it has found relatively few applications so far (mainly in Japan). Real-time control consists of reducing scheduled pollutant emissions whenever dangerous pollution levels are forecast for the near future. Hence, it is a promising alternative management approach in cases where permanent treatment would involve very high costs and/or low effectiveness.

The book is subdivided into two parts, corresponding roughly to the two goals outlined above. Part One considers the role of mathematical models in air-quality planning and includes: a presentation of a decision maker's viewpoint (Dreissigacker et al.); illustrations of various types of models (descriptive and/or decision models) available to decision makers (Dennis, Anderson, Fortak, Olsson, Gustafson and Kortanek); assessments of the role of models in actual decision making (Dennis, Miller); and two papers on the more traditional question of the significance and range of application of descriptive models, i.e., of models that represent the physics of the air-pollution phenomenon (Benarie, Szepesi).

The core of Part Two of the book, which is devoted primarily to real-time control, is a presentation of the IIASA case study of the Venetian lagoon. This research was concerned with the problem of how pollution forecasts can be made, what real-time control actions can be taken on the basis of such forecasts, and what results ensue in terms of costs and effectiveness. Four papers describe various aspects of this research: a predictive wind model (Bonivento and Tonielli); a real-time pollution-concentration predictor (Runca et al., Fronza et al.), which also uses as inputs the wind forecasts supplied by the predictive wind model; and a real-time control scheme (Melli et al.), based on the concentration forecasts given by the real-time concentration predictor, and an evaluation of the cost-effectiveness performance of such a control scheme. Part Two concludes with papers on alternative concentration predictors (Bolzern et al., Soeda and Sawaragi); and descriptions of implementations of real-time forecast and control (mainly alarm) schemes in Japan (Soeda and Omatu) and Italy (Gualdi and Tebaldi).

G. Fronza and P. Melli

CONTENTS

Part One

Models for Air-Quality Planning

AIR-QUALITY MANAGEMENT: REGULATORY PROCEDURES IN THE FRG

H.L. Dreissigacker, F. Surendorf, and H. Weber
Bundesministerium des Inners, Bonn (FRG)

1 INTRODUCTION: THE SIGNIFICANCE OF AIR-POLLUTION STANDARDS

Air-quality standards and emission standards are implemented in the Federal Republic of Germany on the basis of the "Law for the Prevention of Harmful Effects on the Environment caused by Air Pollution, Noise, Vibration and Similar Phenomena" (official translation)–the Federal Air-Quality Control and Noise Abatement Act [Bundes-Immissionsschutzgesetz (BImSchG)].

This law is an extension and modernization of the Commerce and Industry Act of 1869 (Gewerbeordnung) with respect to air-pollution control and is thus based on more than 100 years' experience and tradition in judicial and administrative praxis.

Paragraph 1 of the BImSchG, which defines the purpose of the law and thus its goals, means the following in the area of air-pollution control.

1. Men, animals, plants, and other objects must be protected against those pollutants, which potentially may cause danger, considerable disadvantages, or considerable nuisance because of their nature (properties), extent (volume, significance), and duration (periods of emission, lifetime in the environment).
2. Preventive measures must be taken against the generation of air pollutants.

Installations with a particular potential to cause harmful effects on the environment require a license. For the licensing of these installations the Federal Government has implemented air-quality and emission standards by the 1st General Administrative Regulation under the BImSchG, dated August 28, 1974, as "Technical Instructions for Air-Pollution Control" [Technische Anleitung zur Reinhaltung der Luft (TA Luft)].

These administrative regulations can only be imposed after a hearing of the competent social groups concerned and in agreement with the Federal Council [State Chamber (Bundesrat)].

The air-quality and emission standards are of vital importance for installations subjected to licensing. They form the basis for any administrative activity in all the licensing procedures provided by the law. Plant operators can take account of them when planning and deciding on investments. However, the standards also provide assurance to both those in the vicinity of installations and the public in general that the environment is being protected against harmful effects from air pollution.

1.1 Air-Quality Standards

Air-quality standards must protect the environment against the harmful effects of man-made pollutants.

At present there is a need for air-quality standards for about 300 substances and their chemical compounds. This, however, would require decision makers to have at their disposal elaborate criteria on which to base such standards.

We shall define a criterion as the relation between the *extent* to which an object is encumbered by a harmful substance and the *significance* of the harmful effects to which that object is consequently subjected.

The TA Luft 1974 could only provide certain criteria for sedimentary dust, fine particles, chlorine, hydrogen chloride, hydrogen fluoride, carbon monoxide, sulfur dioxide, hydrogen sulfide, nitrogen dioxide, and nitrogen monoxide.

The criteria that form the basis of air-quality standards must be oriented towards both prevention of and provision against the following: dangers to human health, including the health of those already affected (e.g. bronchitics) and the health of susceptible groups (e.g. pregnant women, babies); major nuisances for humans (e.g. bad smells); and major damage (e.g. injury to property such as animals, plants, etc.).

Air-quality standards have no meaning just as numbers but only in connection with defined measuring and evaluation procedures for the assessment of air quality. Such procedures are also prescribed in the TA Luft. They are based on the experience of long-term measuring programs and ensure that measurements are carried out in such a way that comparison of the results with air-quality standards gives sufficiently safe evidence and that the vicinity and the public in general are protected. These procedures are also thought to limit the costs of measurement programs. In deciding on criteria for air-quality standards the following factors must be considered.

The criteria used in the FRG for individual substances are based on the work of the World Health Organization (WHO), NATO–CCMS, the European Community (EC), and the Air-Pollution Control Commission of the Association of German Engineers (Verein Deutscher Ingenieure). There are no well-defined boundaries between harmful and nonharmful effects: these blurred areas may add up during the elaboration of the criteria, which therefore do not correspond to the air-quality standards defined by the BImSchG either qualitatively or quantitatively. Where the criteria differ quantitatively from the BImSchG air-quality standards, the latter include a safety margin. Thus the blurred areas in the criteria are not allowed to be transmitted to the air-quality standards. Air-quality standards mean strict limits.

Air-quality standards are normally defined as a pair of standards: one for the long-term effects and one for short-term effects. The latter particularly takes into account the fact that peak concentrations even during short periods can lead to harmful effects.

On September 6, 1978, the Federal Government decided by an amendment of the TA Luft on additional air-quality standards for sedimentary dust and fine particles of lead and cadmium. Furthermore, a standardized transmission model was to be prescribed to allow the forecast of the changes in air quality caused by projected new installations or by alterations of existing installations. On the same day the Federal Government also decided on an ordinance concerning the annual declaration by the operator of the emissions from an installation and on an administrative regulation concerning regional emission inventories to be instituted by the competent authorities. The instruments for local and regional air-quality management provided by the BImSchG have thus been further extended.

1.2 Emission Standards

Because of the requirements of the BImSchG cited earlier, rather rigid limits are set for economic considerations during the establishment of air-quality standards. This is not true to the same extent for emission standards, which are based on the latest "state of technology".

The TA Luft 1974 defined the "state of technology" as follows:

> "When judging whether–considering the special circumstances of the particular case–the actual state of technology can be achieved, advanced comparable processes and plant installations proven by operation are to be considered."

The requirement "proven by operation" was a static element in this definition, impeding the progress of technology. There were in fact considerably different conditions for the innovation of environmental and productive technologies: environmental technologies first had to be demonstrated to be proven by operation, whereas productive technologies were normally converted into economic plant size after pilot-plant demonstration using scale-up factors.

The legislature, drawing conclusions from this unsatisfactory situation, defined the state of technology in accordance with general economic practice as follows (BImSchG 1974).

> "The latest state of technology in the sense of this law shall mean that state of development of advanced processes, facilities, or operation methods, at which a method of controlling emissions appears certainly to be suitable for practical application. In determining the latest state of technology comparable processes, facilities, or operation methods, which successfully have been proven by operation, shall particularly be considered."

For the purposes of air-pollution control the TA Luft 1974 made this legal definition precise as follows.

> "Advanced methods for controlling emissions, which have been proven by operation, shall particularly be considered to be the latest state of technology. Methods which are not conclusively proven by operation for the respective application may be considered to be the latest state of technology in substantiated cases, e.g. if they are proven to such an extent that the projected application is feasible without an unreasonable risk. Comparable

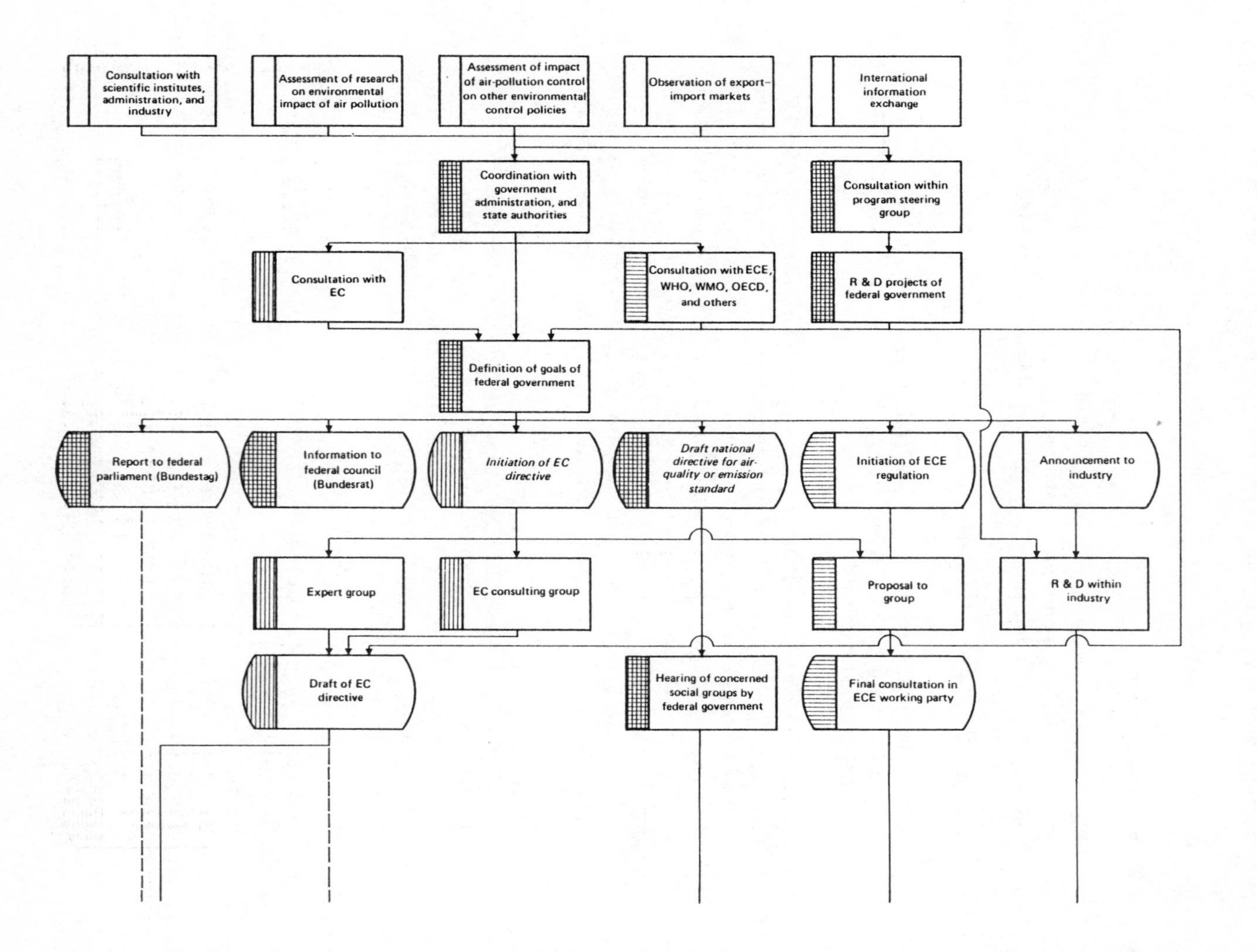
Consultation with scientific institutes, administration, and industry
Assessment of research on environmental impact of air pollution
Assessment of impact of air-pollution control on other environmental control policies
Observation of export–import markets
International information exchange
Coordination with government administration, and state authorities
Consultation within program steering group
Consultation with EC
Consultation with ECE, WHO, WMO, OECD, and others
R & D projects of federal government
Definition of goals of federal government
Report to federal parliament (Bundestag)
Information to federal council (Bundesrat)
Initiation of EC directive
Draft national directive for air-quality or emission standard
Initiation of ECE regulation
Announcement to industry
Expert group
EC consulting group
Proposal to group
R & D within industry
Draft of EC directive
Hearing of concerned social groups by federal government
Final consultation in ECE working party

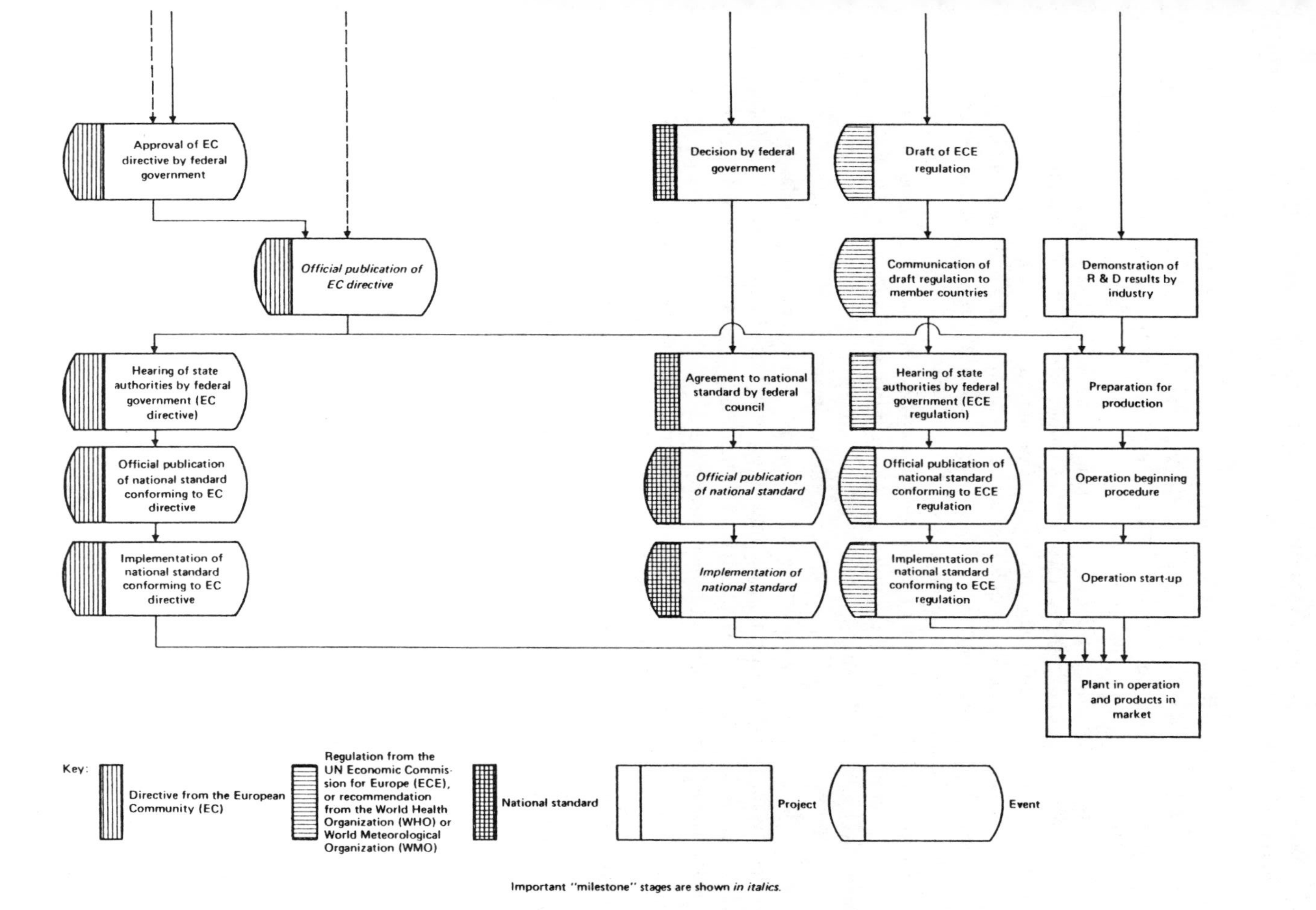

FIGURE 1 Flow diagram showing the standard procedure used in the FRG for the elaboration and implementation of air-pollution control standards.

operation methods, facilities, or processes shall particularly be considered when determining the latest state of technology for the respective application. If such operation methods, facilities, or processes do not exist, particularly stringent requirements shall be applied to the judgement of whether a method of controlling emissions appears certainly to be suitable for practical application."

The legal definition and the way in which it is made precise by the TA Luft 1974 remove the earlier-mentioned impediment to the progress of air-pollution control technology while protecting the plant operator against undue risks in the transfer of recently developed processes to industrial application.

In the implementation of the emission standards according to the latest state of technology the principle that the required means should remain within due limits has to be observed. Excessive requirements are thus not permitted; i.e. generally binding emission standards which would call in question the economic basis of a branch of industry are not tolerable. Emission standards are based in general on the results of measurement programs carried out in modern installations.

The significance of important single sources for regional air-quality management is taken into account by raising the emission control requirements with increasing size of installations.

Administrative regulations like the TA Luft are obligatory on the competent authorities which may deviate from these obligations in substantiated cases only. Subsequent requirements for plants in operation are limited to certain conditions defined by the TA Luft, except in those cases where the vicinity or the public in general is not sufficiently protected against harmful effects to the environment.

The emission standards include a certain safety margin allowing for possible fluctuations of plant operation. This is necessary since frequently exceeding the emission standards would mean a violation of the law and this would lead to severe fines. This safety margin is also important where plant operation is automatically stopped when emission standards are exceeded. Standards that were too restrictive would lead to too-frequent plant shutdown.

1.3 Conclusion

The air-quality and emission standards set forth in the TA Luft allow industry to assess the decisions of the licensing authorities and thus to have an assured basis for planning and deciding on investment.

2 THE ELABORATION OF TECHNICAL–SCIENTIFIC STANDARDS FOR AIR-QUALITY CONTROL

Technical–scientific air-quality control standards are elaborated within the framework of national legislation and international obligations. The process thus follows, with some variations, a certain standard procedure which is illustrated by the flow scheme in Figure 1. The figure shows the process of national and international rulings, the latter as far as they are of importance for the FRG in cases where their results must be included in national legislation. This is particularly true for the European Community (EC) directives and, to a certain degree, for the UN Economic Commission for Europe (ECE) recommendations. The process starts when impulses are given for the elaboration of a standard, for example from

- consultation with scientific institutes, administrations, and industry;
- assessment of research on the environmental impact of air pollution;
- assessment of the impact of air-pollution control activities on other environmental-pollution control policies;
- observation of export and import markets; or
- international exchange of information and experience.

The process ends with the implementation of standards and the operational start-up of the installations concerned or the introduction to the market of the product concerned.

The network of manifold interrelations between the individual activities has been reduced and simplified in order to give a clear picture.

THE IMPACT OF MODELS ON DECISION MAKING: AN ASSESSMENT OF THE ROLE OF MODELS IN AIR-QUALITY PLANNING

R.L. Dennis
Environmental and Societal Impacts Group, National Center for Atmospheric Research, Boulder, Colorado (USA)

1 INTRODUCTION

Models are used in all aspects of air-quality planning where prediction is a major component – from episode forecasting to long-term planning. As measurement technology, computer technology, and knowledge of the atmosphere and the processes that take place in the atmosphere have been improved, these models more and more have become computer models. In this paper the term "model" will only refer to computer models.

Imperfect as the models may be for the questions that are asked of them, they are the best and usually the only means of making the predictions needed. When built and used properly, models require that assumptions be laid out more explicitly than when decisions or planning take place without the use of a model. This more-rigorous formulation of assumptions is a great advantage for decision making, even if great uncertainty exists in the model. It appears that the benefits of models have been gradually recognized as greater than the disadvantages that arise from their imperfections.

In the United States, air-quality models are having a significantly increasing impact on decision making. Some of the major underlying causes for this are the Clean Air Act Amendments of 1977. While the Clean Air Act of 1970 has been described as a major piece of innovative social legislation (Ingram, 1978), the Clean Air Act Amendments of 1977 and the rules and regulations promulgated on June 19, 1978 concerning the prevention of significant deterioration of air quality are more significant for the modeling community. It was in these latter pieces of legislation that the use of air-quality models was made mandatory both before permits for new sources of pollution were issued and for the formulation of the state implementation plans for air-quality maintenance. The use of models in decision making is increasing not only in the United States but also in many other countries such as Sweden (Schütt and Bergman, 1979), the Federal Republic of Germany (Schulz and Stehfest, 1978), and the German Democratic Republic (Foell, 1979). In the United States, however, the question is becoming not only one of the magnitude of the impact of models on decision making but also of the quality of this impact.

In this paper we will discuss the quality of the impact that air-quality modeling has on related decision making in terms of the roles played by uncertainty (uncertainty in the model prediction due to the limitations of the model in simulating the physical processes of dispersion) and by model formulation (the way a model is structured or used in order to answer specific types of questions). Both short-term and long-term planning will be discussed, with more emphasis being given to regional planning. Episode planning and short-term management are included under short-term planning.

2 SHORT-TERM PLANNING

Short-term planning involves (a) the decisions required to regulate or control the system of emissions that already exists and (b) the decisions required to regulate and control actual incremental changes in the system of emissions (the addition of new sources, the modification or removal of old sources). Control of the former is discussed under episode planning, while control of the latter comes under short-term management.

2.1 Episode Planning

Episode planning is like weather prediction in that the time scale is similar (1–3 days advance prediction) and the uncertainty in the prediction is tied to the same phenomenon. However, unlike weather prediction episode planning is concerned only with extreme events (i.e. when standards are expected to be violated).

Studies have shown that numerical forecasts improved dramatically between 1950 and 1970 but that progress has slowed to give very small improvements since then (Shellman, 1977; Sanders, 1979). At present, Man is still a better forecaster for periods of up to two days, but machine and Man are about equal for third- to fifth-day forecasts. There are also important indications that Man may be better than numerical models at predicting extremes (A. Murphy, personal communication).

As mentioned above, it is the extremes that are important for episode planning. The estimated annual cost of calling air-pollution alerts in Chicago is between \$6 million and \$10 million. The cost increases dramatically as one goes from the first stage to the fourth stage (emergency); this last stage is anticipated to cost \$36 million (Cohen et al., 1977).

It is interesting to note that regional air-quality models were extensively used in order to set up the stages of the "episode emergency plan". Because of the cost involved and the potential economic disruption it was very important to be able to estimate the effects of reductions in various emissions relative to the costs incurred, so that the least-costly measures could be taken in the first stages and the most-costly measures in the last stage or emergency alert. The episode planners worked with the longer-term regional models in order to define each step and to be able to explain and justify measures taken at each decision point. Certainly emergency alerts cannot be ad hoc affairs.

Weather forecasting experience suggests that relying only on the model might be more costly than relying on Man's judgment because Man predicts the extremes better. Man is able to integrate additional factors into forecasts from personal experience and thus to reduce some of the uncertainty inherent in the numerical prediction. For episode prediction it appears that a close mix of Man and machine can provide the best-quality prediction. One policy question is whether the person involved receives a reward for reducing the uncertainty that is sufficient to offset the risks involved in an incorrect prediction. Without sufficient reward the forecaster may prefer to let the machine take the blame for an incorrect forecast and to let someone else assume the added annual cost that might arise from relying only on the numerical prediction.

2.2 Short-Term Management

Included under short-term management are new-source permit evaluation and near-term regional planning, i.e. the state implementation plans required to demonstrate maintenance or improvement of air quality for the years 1982 and 1987. These two decision areas are the areas where air-quality models are most intensively used and have the greatest impact in the United States. They are also the areas that give rise to most of the arguments about air-quality models and the most heated debates about basing decisions on these models.

The overriding impact on decision making, due to the use of air-quality models made mandatory by the Clean Air Act Amendment of 1977, is the qualitative fact that we have been forced into learning how to make decisions based on predictions from simulation models. The consequences of how we develop a decision-making process that includes predictions from simulation models will be long lasting. What extra tools we develop to help to bring the use of scientific information into the decision-making process may be far more important in the long run than the fact that Simulation Model X was used to help make Decision Y. This will also be true for long-term planning.

Thus many of the arguments center on how to arrive at a decision when the uncertain prediction comes from a computer rather than a man. One cannot harangue a computer model or use the art of persuasion in the way one can with a person. Other things can be done, however, with the help of technical experts. Given the uncertainty, how does one control, for example, the subtle influencing of the model predictions for a particular interest group? The learning that is involved in adapting the use of scientific information based on simulation modeling to the present decision-making process is most evident in the areas of new-source permit evaluation and near-term regional planning.

2.3 Permits to Operate New Sources of Pollution

In evaluating the effects of a new pollution source the uncertainty of the prediction of the air-quality model is greater than the uncertainties of most other elements involved in making tradeoffs for the decision or in making a decision whether or not standards will be violated. While economic costs and engineering data are well known compared to the air-quality prediction, effects such as human health impacts or visibility degradation are more poorly understood than the air-quality prediction. Thus it can be seen that arguments about the uncertainties of an air-quality model, in terms of its inputs and parameterization, frequently spill over into questions about the uncertainty of the effects, since consideration must be given to the possibly large costs required to achieve certain reductions in emissions. This is particularly true of the Electric Power Research Institute arguments concerning SO_2 scrubbers (C. Comar, personal communication; Mirabella, 1979) and the arguments of the automotive industry concerning CO emission standards (Anon., 1979).

Decisions made on new-source permits are not usually reversible. If a scrubber is not installed and it is later found that it should have been (or vice versa) it costs a lot more to change than the original investment and most probably it will not be changed. An incorrect decision can cost money, and we must live with the decision. However, the cost of an incorrect decision taken now is most likely to be less than the cost of an incorrect decision taken ten years from now. Meanwhile a great deal is being learned about the use of air-quality models in decision making.

For example, the pressure to use the models of the Environmental Protection Agency (EPA), even though they may be inferior to some others, may have significant results because these models are public and both parties are required to use the same model. When the Northern Cheyenne Indians used the EPA valley model they got a different answer from that obtained by Montana Power and Light regarding whether the coal strip plant outside the reservation met standards on the reservation. The meteorologist for Montana Power and Light thought that the EPA model was not properly parameterized for the terrain and therefore adjusted the meteorological frequencies in the various categories that are input to the model. With the adjusted inputs, it was predicted that the coal strip plant would meet the standards. With the original inputs used by the Northern Cheyenne, it was predicted that the plant would not meet the standards.

Had the models not been public and had both parties not used the same model, the dispute could not have been narrowed down to the relevant uncertainties affecting the predictions. One of the powers in using a model as an aid to the decision making, i.e. the ability to pinpoint the differences in assumptions, would have been lost because the model requires that the assumptions be made explicit. When trying to improve decision making, it is undesirable to have the assumptions unspecified and to discuss only the results from two different models. In addition, the use of the same model by both parties puts the model through its hardest tests and points out its most-significant inadequacies, giving an impetus to improve the model. Because of the large reaction to the inadequacies in their models, the EPA are now starting intensive research on terrain effects. If the models were not so directly involved in decision making it is doubtful whether their improvement would occur so rapidly.

By testing and correcting the models now, while the costs of a wrong decision are more bearable, we will improve the quality of the input from air-quality models in the future when this will be really necessary to ensure proper decision making. It is essential not to wait until the crisis is here before starting to use models because, as will be discussed in the next section, it takes time to learn how best to use the models.

Meanwhile the lesson of the Northern Cheyenne example should not be lost. We must not succumb to the sorcery of the computer and let the assumptions and inputs to the model, or the model itself for that matter, remain hidden behind a veil of obscurity. The usefulness of a model for careful, rational decisions will be immeasurably enhanced by greater exposure and documentation of the model.

2.4 Near-Term Regional Planning

The air-quality models used in near-term regional planning are usually large and data intensive. Regional planning is defined here to be urban and regional planning, i.e. planning for large (or small) metropolitan areas and their contiguous surroundings. Larger regions, such as a country or a US state, are not included in this discussion. However, in this type of planning the air-quality model becomes one of several large models and the problem becomes one of dealing with uncertain predictions from a set of models, although that set of models may have little or no organized superstructure. In this case the uncertainty of most of the inputs to the air-quality model is as large as the uncertainty in the air-quality predictions. Regional planning also moves away from the air-quality standards as the only focus of the decision making, although it must be remembered that maintenance or meeting of the standards is the driving force behind the modeling and decision making.

The factors that appear to affect the quality of the impact of the models on decision making are (1) how well the models are used for predicting the relative change produced by different options, (2) the amount of communication and learning allowed between model, analyst, and decision maker, and (3) how clearly social values and uncertainty are treated in the analysis.

Air-quality models are playing a very important role in the newly-required emphasis on air quality in regional planning. While air-quality models have a degree of uncertainty in their *absolute* predictions they do a more credible job in terms of *relative* changes. This is their present strength and the reason for their large impact on decision making. It is doubtful that this situation will change before the completion of the next round of planning for 1987. Most users therefore obtain the most valuable output from air-quality models when they consider relative changes.

This implies that a large amount of time must be spent in calibrating and using an air-quality model before it will start being a tool and a guide in planning options. For example, in the South Eastern Wisconsin Regional Planning Commission (SEWRPC) this process took four years. The first year was the most discouraging. There were problems in integrating the regional-planning model with the air-quality model. Changes that looked significant from the planners' perspective made little or no difference in air quality. It was a learning process. The next two years were spent in calibrating, in smoothing out the communication links, and in starting to learn how to dissect the problem to make the best use of the air-quality model.

The fourth year was a success. The system was calibrated. The SEWRPC knew how to take the system apart, examine the relevant piece, and then put the pieces back together. They learned how to use the air-quality model as one of their tools and how to develop options that did make a difference in air-quality prediction. During this time the SEWRPC spent at least $30,000 on model runs and the various air-quality models were run more than 500 times. This is what is required before air-quality models will have a real impact on planning and decision making in a regional context. It appears that the San Francisco Association of Bay Area Governments had an experience somewhere between that of the SEWRPC and that of the Denver authorities described immediately below (Feldstein and Tranter, 1979).

The introduction of a model through consultants and a few quick studies does not meet the needs of decision making because it does not allow the uncertainties in the air-quality models to be understood and accounted for in the decision-making process. The consulting study commissioned by the Denver Regional Council Of Governments (DRCOG) and the Air Pollution Control Division (APCD) of the Colorado Department of Health for an air-quality analysis raised more questions than it answered and pointed out a group of uncertain inputs that should be looked at. Because of cost, only two days could be modeled, one in summer and one in winter. However, the cost of repeating the study as well as the cost of defining the emission inputs for the air-dispersion model ($5000 per run for the transportation model) was too great. Thus the DRCOG and the APCD have not (at the time of writing) yet really obtained the benefit of the air-quality models; nor will they in future if they have to rely on consultants for the air-quality analysis. They are still at the stage with the modeling that the SEWRPC reached three years earlier, finding that air quality is more difficult to influence than had been imagined. Because they did not have an air-quality modeling group within their office or at nearby universities and laboratories, the DRCOG and the APCD did not have a convenient or inexpensive way to ask lots of questions in order to learn how best to use the models for their needs.

The author has done a preliminary analysis of one of the policy options that the DRCOG is considering to improve the air quality of Denver. It took 70 runs of the dispersion model to achieve a preliminary analysis; 15 of the runs were the preliminary analysis itself and the rest were for calibration and learning. It is expected that a couple of hundred runs will be necessary before the analysis is complete. Given the present state of the art, just having a model available for a limited number of studies does not mean that the air-quality model will have a significant impact on regional decision making. The uncertainty inherent in the models dictates that the models should be continually available and in use for some time before they will mesh in smoothly as a tool for decision making.

In this type of complex modeling and decision-making process each discipline points to the uncertainties in the other parts of the analysis as being responsible for the larger part of the difficulty in the analysis. For example, air-dispersion analysts point to the uncertainties in the damage functions as being very important because so many people put such a high value on human health impacts, whereas their colleagues who work on dose–response functions point to air-quality measures and the uncertainty of exposure as the major problem (Morgan et al., 1978; Spengler, 1979).

Because the models have a degree of uncertainty in their predictions, because the predictions of the model may be different from what we intuitively would have guessed, because no model covers all aspects of the problem, and because no model has been so thoroughly validated as to be without questions, decision makers are not going to (or at least should not) accept blindly what comes out of the computer. Therefore the most successful uses of models in decision making have involved a high degree of effort in communication between modeler and decision maker. In the SEWRPC work this communication was developed through the actual participation of the modelers in the planning meetings, through the constant availability of the modelers to answer questions, through the mutual consensus that everything should be documented as carefully as possible, and through the willingness on both sides to treat documentation seriously.

In the Chicago episode work, though this was economics modeling more than air-dispersion modeling, the communication occurred throughout the study in the form of informal policy dinners. Policy planners and legislators were continually kept abreast of developments and their opinions were solicited in an informal relaxed manner. This was deemed to have contributed greatly to the success of the study (P.E. Graves, personal communication; Holcomb Research Institute, 1976). In the San Francisco Bay Area both types of communication were used (Feldstein and Tranter, 1979).

Although the models will improve with time, the need for communication will not change. A certain degree of uncertainty will always exist. Therefore an important area of work that needs to be developed and which will probably greatly improve the use of all models in decision making is the development of methods for explicitly handling uncertainty in the air-quality models vis-à-vis social values. The impact of models can be improved, for a given degree of uncertainty in the models, if mixups between differences in numbers and differences in social values can be disentangled and delineated. Arguments over social values are often confused with arguments over numbers. Given a set of social values, the uncertainty in the numbers may not in fact change the decision anyway, or the social values may provide a completely different perspective to that given by the numbers (Mumpower and Dennis, 1979).

In addition, near-term regional planning needs to expand beyond its present use of air-quality models, and ask new and different questions of the models. Because we must now project general economic and human behavior, the inputs required by the air-quality models are more uncertain than the air-quality predictions. Concentration only on reducing the uncertainty in the air-quality models may not provide a better answer. We may need to ask a different type of question of the models for their impact to be enhanced. This aspect of reformulating the questions and even the air-quality models themselves is discussed in the next section on long-term planning.

3 LONG-TERM PLANNING

Most if not all of the comments made in Section 2.4 for near-term regional planning apply to long-term regional planning (i.e. 10–50 years into the future). Several problems that begin to appear in near-term regional planning are seen more fully in long-term planning. Most of the uncertainties now encountered are found to be larger than the uncertainties in the air-quality predictions. Air quality becomes part of a mosaic of issues – one of many tradeoffs and many uncertainties. Social values are more important than ever and must be explicitly taken into account in as formal a way as possible. The variety of policy questions to be asked of the air-quality models is larger; therefore the models must be reformulated and simplified for quicker and more general answers. Nevertheless, the power of such simplified models may still be large because subsequently they can be reformulated to a different level of robustness and can thereby reduce some of the uncertainty in prediction.

We are still very much in the learning phases of long-term planning. Many of the near-term regional models serve as springboards for long-term planning models such as the Lower Delaware Valley study by Resources for the Future (Spofford et al., 1976; Holcomb Research Institute, 1976). On the basis of experience to date, three areas of investigation show promise in improving the impact of models on long-term decision making. They are: (1) formal linking of the technical and social value assessments, (2) reformulation and/or simplification of models so that they become more policy-relevant, and (3) explicit attention to the uncertainty in the model predictions in the framework of the decision-making environment. All of these presume close communication between modeler and decision maker.

3.1 Linkage of Technical and Value Assessments

Long-term planning is indistinguishable from social policy formation. Social policy formation in this context requires integration of two types of concerns: (a) social values concerning "what should be accomplished" and (b) technical assessments concerning "what it is possible to accomplish". Those involved in technical assessment are realizing that technical information needs to be integrated with social values if models are to be used effectively (Dennis, 1979; Ford and Gardiner, 1979; Buehring et al., 1976). The integration of social values with technical information constitutes a cognitive problem of considerable difficulty (Mumpower et al., 1979). Quite commonly either (a) policy makers are required to interpret difficult technical information, thus being required to become amateur scientists or engineers, or (b) technical experts are asked to make the policy decisions, thus being required to become proxy policy makers.

The Symmetrical Linkage System (SLS) has been proposed as a method that at least partially redresses some of the problems commonly found in social policy formation (Hammond et al., 1977a, b; Mumpower et al., 1979; Mumpower and Dennis, 1979). This method prescribes the use of (a) judgment analysis to construct explicit, quantitative models of the social goals and objectives of the policy-making process, (b) technical analysis to construct quantitative models of the scientific information relevant to the achievement of the social goals and objectives, and (c) computer algorithms to link together these two types of quantitative models and thus to identify policies that satisfy the social goals and objectives as fully as possible. The method is described as symmetrical because equal attention is given to both social values and scientific information.

The SLS approach was originally developed and applied in a pilot study in the context of a university faculty planning problem (Hammond et al., 1977a). It has also been applied on an illustrative basis to an analysis of the national energy policy of the United States (Hammond et al., 1977b). Most recently the SLS approach has been used in a pilot study for policy analysis concerning the regional air-pollution management problem of Denver (Mumpower et al., 1979). This pilot study has developed into a full-scale investigation of the air-management problem of Denver carried out by the National Center for Atmospheric Research and the Institute of Behavioral Sciences, University of Colorado.

3.2 Reformulation of Models

Long-range planning and option analysis are constrained to a detrimental degree by their present dependence on the same air-quality models as are used in near-term regional planning. Too much preparation of the input information and too much sifting of the output is required. The point is not that these models should be replaced but that they should be preceded by a set of simpler models that are easy to parameterize and run for widely different options. Such simplified models provide a broad exploration as well as a preprocessing of the various policy options. They also profile the most critical parameters or responses of the system in terms that are relevant to policy decisions, providing a more lucid and easily comprehensible description. In addition, they can highlight the robust features of the system, i.e. the features that are useful as policy indicators because they are less sensitive to uncertainties in inputs.

These simplified models are not "simple minded" but are based on reformulations of the more detailed models or on sound theoretical principles. The relationship between the simplified models and the more detailed models should be like the principle of correspondence in physics between classical mechanics and quantum mechanics. One example of a detailed model reformulated into a simplified policy model is the Smeared Concentration Approximation (SCA) method (Dennis, 1978). This method, developed for non-reacting chemical species with no strong diurnal variation in concentration, defined the spatially averaged ground-level concentration for an urban area as the most appropriate policy indicator of air pollution for energy/environmental analysis because the social values involved emphasized human health impacts. This indicator was shown to be very robust with respect to almost all the details of the location of emissions within the urban area. Thus the need to worry about the details of any land-use environment in the long-term analysis disappeared, and the need to consider the uncertainty was removed.

Initial studies have shown that the SCA method can provide very useful policy-relevant air-quality analyses both at the urban level and at the national level. It provided the policy analysts of the German Democratic Republic with a generalized evaluation of the benefits of district heating for air quality in an urban area (Foell, 1979). The analysts are now pursuing a more detailed evaluation for their own decision makers (W. Kluge, personal communication). The SCA method also provided the Austrian Ministry of Environment with a generalized analysis of the efficacy of several different possible SO_2 emission standards in reducing effects on human health due to SO_2 air pollution (Foell et al., 1979).

At present this type of model reformulation and simplification for long-term planning analysis is being pursued for the modeling of pollutants with a strong diurnal concentration cycle (e.g. CO) and for oxidants. This work is taking place in the context of an applied study of regional air-quality management for Denver.

3.3 Uncertainty within the Decision-Making Environment

Very little work has been done to address uncertainty in the context of the social value structure. Although techniques of decision theory are capable of accounting for uncertainty, this is seldom if ever done, reliance being placed only on expected values. However, the social value structure articulated by a decision maker may give very different weights to the importance of different attributes of a problem. Thus, when all the attributes of the problem are combined into a judgment of desirability or preference score, a variable with a large scientific uncertainty might produce only a small range of variation in the decision maker's preference score. It is therefore felt that a description, for a given policy, of the effect on the decision maker's desirability score due to one, several, or all of the contributing uncertainties in technical assessments would be a meaningful framework for presenting the effect of uncertainty to the decision maker.

Work has been started in this direction (Mumpower and Dennis, 1979). The lowest desirability score in the range is defined as the score that results when all the variables are moved to the uncertainty limit in the direction of lower desirability. The highest desirability score is defined by doing the opposite. The mid-range score of desirability is defined by all the variables taking their expected value. The uncertainties are dealt with simultaneously rather than on a dimension-by-dimension basis. Thus different scenarios or options can produce different ranges in the desirability scores.

All policy alternatives for which the minimum overall desirability score is exceeded by the maximum overall desirability score of some other policy alternative are identified as non-dominated alternatives. In general, it is found that the set of non-dominated alternatives tends to become larger. It is possible that the desirability score for the expected value of alternative A will be less than that of alternative B but that the range of the desirability score of alternative A will be much larger than the range for B. Thus if the decision maker believes that a key variable will move away from the predicted expected value towards an uncertainty limit that increases the desirability score for alternative A (e.g. an expected decline in fertility that is larger than predicted), then that decision maker may want to consider risking the choice of alternative A over B even though the desirability score based on expected value is larger for B than for A. This is a different type of consideration from Herbert Simon's concept of satisficing.

Many modelers and systems analysts require a better accounting of uncertainty, but if all we do is spread out expected values and produce a widening swath of numbers over time we are not doing the decision maker a service. Without a cogent and relevant analysis of uncertainty the decision maker is once again forced into the role of amateur scientist as he/she tries to make sense of an even larger array of numbers, or the scientist is once again asked to become an amateur policy maker. The analysis of uncertainty in terms of its effect on the social desirability rating is one proposed method of addressing the difficulty of communication between modeler and decision maker about uncertainty without reversing the roles of analyst and policy maker. Putting the uncertainty analysis in the framework of social values should be a step forward.

4 CONCLUSIONS

For new pollution-source permits (short-term planning) and near-term regional planning this paper should have been entitled "the impact of decision making on modeling", instead of the other way around. In the United States experience, the Clean Air Act Amendments of 1977 have played a significant role in pushing the use of modeling to the foreground. Because the use of scientific information in decision making will inherently have uncertainty associated with it, this process of integrating modeling into decision making for air-quality management would have proceeded much more slowly without such a legislative push.

Much is being learned about how to make better use of models. Models are influencing decisions. Mistakes will be made but the process of using models has been found to be realistic and reasonably flexible. This is still a learning phase but it is far better to learn rapidly now, when the costs of mistakes are lower, than later, when there will be bigger crises and the costs of mistakes will be higher. Energy decisions of enormous magnitude that could affect air-quality options for the next 30 years will be made in the next ten years. The sooner the use of air-quality models can be put on a firmer, more-routine basis and can be interjected into such decision making, the better.

The long-term planning and analysis capability of air-quality modeling must improve before this will happen, however. The more detailed work of near-term planning and analysis can serve as a springboard for this reformulation but we must branch away creatively from the large detailed models in order to have a substantial and effective impact on decision making. Three avenues proposed for this branching are (1) learning to reformulate the models and to ask different questions of them, (2) linking the technical analysis with quantitative social-value analysis, and (3) treating uncertainty in the model prediction vis-à-vis the decision process and the social values involved.

REFERENCES

Anon. (1979). How to figure the cost of living a longer life. Technol. Rev., back cover.

Buehring, W.A., Foell, W.K., and Keeney, R.L. (1976). Energy/environment: application of decision analysis. RR-76-14. International Institute for Applied Systems Analysis, Laxenburg, Austria.

Cohen, A.S., Macal, C.M., and Cavallo, J.D. (1977). Cost-effectiveness analysis of the Illinois ozone episode regulation. In Environmental Pollutants and the Urban Economy. Argonne National Laboratory, Energy and Environmental Systems Division and Center for Urban Studies, University of Chicago, Chicago.

Dennis, R.L. (1978). The smeared concentration approximation method: a simplified air pollution dispersion methodology for regional analysis. RR-78-9. International Institute for Applied Systems Analysis, Laxenburg, Austria.

Dennis, R.L. (1979). The role of modeling: a means for energy/environmental analysis. In G.S. Goodman (Editor), Impacts and Risks of Energy Strategies: Their Analysis and Role in Management. Academic Press, New York.

Feldstein, M. and Tranter, R.A.F. (1979). Anatomy of an air quality maintenance plan. J. Air Pollut. Control Assoc., 29:339–364.

Foell, W.K. (Editor) (1979). Management of Energy/Environment Systems: Methods and Case Studies. IIASA International Series, Vol. 5. Wiley, Chichester, UK.

Foell, W.K., Dennis, R.L., Hanson, M.E., Hervey, L.A., Hoelzl, A., Peerenboom, J.R., and Poenitz, E. (1979). Assessment of alternative energy/environment futures for Austria: 1977–2015. RR-79-7. International Institute for Applied Systems Analysis, Laxenburg, Austria.

Ford, A. and Gardiner, P.C. (1979). A new measure of sensitivity for social systems simulation models. IEEE Trans. Syst., Man, Cybern., 9:105–114.

Hammond, K.R., Mumpower, J.L., and Smith, T.H. (1977a). Linking environmental models with models of human judgment: a symmetrical decision aid. IEEE Trans. Syst., Man, Cybern., 7:358–367.

Hammond, K.R., Klitz, J.K., and Cook, R.L. (1977b). How systems analysis can provide more effective assistance to the policy-maker. RM-77-50. International Institute for Applied Systems Analysis, Laxenburg, Austria.

Holcomb Research Institute (1976). Environmental Modeling and Decision Making: the United States Experience. Praeger, New York.

Ingram, H. (1978). The political rationality of innovation: the clean air amendments of 1970. In A.F. Friedlander (Editor), Approaches to Controlling Air Pollution. MIT Press, Cambridge, Massachusetts.

Mirabella, V.A. (1979). Atmospheric dispersion modelling: a critical review. J. Air Pollut. Control. Assoc., 29:931–934.

Morgan, M.G., Morris, S.C. Meier, A.K., and Shenk, D.L. (1978). A probabilistic methodology for estimating air pollution health effects from coal fired power plants. Energy Syst. Policy, 2: 287–309.

Mumpower, J. and Dennis, R.L. (1979). The linkage of judgment analysis and technical analysis: a method for social policy formation. In Proc. IEEE Int. Conf. Cybern. Society, (1979).

Mumpower, J., Veirs, V., and Hammond, K.R. (1979). Scientific information, social values and policy formation: the application of simulation models and judgment analysis to the Denver regional air pollution problem. Rep. 218. Center for Research on Judgment and Policy, University of Colorado, Boulder, Colorado.

Sanders, F. (1979). Trends in skill of daily forecasts of temperature and precipitation, 1966–1978. Bull. Am. Meteorol. Soc., 60:763–769.

Schulz, V. and Stehfest, H. (1978). Berechnungsgrundlagen zum Zielsystem des Optimierungsmodells für das Energieversorgungssystem Baden–Württembergs. Rep. 09-03-02 PO3B. Kernforschungszentrum Karlsruhe, Eggenstein–Leopoldshafen, FRG.

Schütt, T. and Bergman, L. (1979). Planning and management of the Swedish energy/environment systems. IIASA Conf. on Management of Regional Energy/Environment Systems.

Snellman, L.W. (1977). Operational forecasting using automatic guidance. Bull. Am. Meteorol. Soc., 58:1036–1044.

Spengler, J.D. (1979). Atmospheric dispersion modeling: a critical review. J. Air Pollut. Control Assoc., 29:929–931.

Spofford, W.O. Jr., Russell, C.S., and Kelly, R.A. (1976). Environmental quality management: an application to the lower Delaware Valley. Res. Paper R-1. Resources for the Future, Washington, D.C.

CASES IN THE APPLICATIONS OF AIR-QUALITY MODELS IN POLICY MAKING

C.G. Miller
Harvard University, Cambridge, Massachusetts (USA)

1 INTRODUCTION

The focus of this research is the decision-making process, how policies and decisions are formed within the various environmental protection agencies in the United States, and what role the air-quality models play within that process. Technical information derived from the models is just one of the many inputs into this decision-making process. Policy formation is a complicated process that involves experience, political insight, and political pressures, as well as technical information.

The complexity of the decision process and the intrinsic uncertainties of the scientific data base combine to make this analysis of model usage a wide-ranging study. Several meanings can be ascribed to the "use of models", depending in part on the motives and organizational or institutional constraints operating on the decision maker. Likewise, the motives and constraints on the modeler developing the technical information may determine what information is available to the decision maker. Thus, this study of air-quality models involves differentiating the uses policy makers might have for the kind of technical information provided by air-quality models and what elements of both the policy and research environment restrict or promote a certain type of use.

Various uses are made of technical information or model results by decision makers. The situation that generally comes to mind is the use of models to solve a problem. A commonplace and important use is the consideration of model results in deciding to issue a permit for a new pollution source or approving a master plan for economic development. Other examples, however, include using technical information in conceptualizing or defining the problem, or capitalizing on an opportunity provided by a basic research project. These uses do not search for a model whose conclusions can be applied to a predetermined, explicit problem or issue. Instead, they apply the results of one or more research efforts, perhaps undertaken for other reasons, to the development of strategies or problems not previously thought to be the priority issues.

Most policy makers and scientists would agree that, to the extent possible, air-quality models should be used in these situations. However, decision makers can use models in other ways which may be less acceptable, especially if their use is divorced from the political context. One case could be the employment of technical information for political advantage such as to support or justify a predetermined position or to delay having to make a decision at all. Similarly, research could be used for self-serving ends by both policy makers and researchers, such as to maintain their prestige or to expand their domain of influence.

This project is concerned with examining case studies of the application of air-quality models within a policy framework. Each study outlines the types of uses that were made of the model results and what motivational and organizational factors played a part in determining how the results were applied or what problems resulted from the attempt to employ a particular model. The case studies explore which aspects of the policy and the research environments are likely to produce certain uses and how the various aspects interact to hinder or promote both appropriate and inappropriate uses.

The usual constraint or limitation one thinks of when trying to describe the environments within which a decision maker or modeler must work is the state of society. For the decision maker limitations exist because of (1) political ideologies as expressed in laws which narrow down the range of actions that can be taken, and (2) budgetary and resource constraints*. For the researcher, the state of the art and the maturity of the theory and methodologies of any particular discipline, set absolute constraints within which one must work.

However, these are not the only factors determining how a model is developed by the modeler, chosen by the analyst, and used by the decision maker. Other factors include background or educational training, previous job experience, professional interests, and institutional and organizational settings. These factors are examined in the case studies.

2 USES OF MODELS

Models can be used in several different ways within a policy-making framework. These uses can be divided into four categories:

(i) models used for problem solving;
(ii) models used for conceptualization;
(iii) basic research leading to application;
(iv) models used for political advantage or to advance self-interest.

These categories overlap with those outlined by Weiss (1978) for social-science research. Similar uses are described by Greenberger et al. (1976).

2.1 Models Used for Problem Solving

The primary use of research by a decision maker is for problem solving: the application of technical information to the problem at hand for the purpose of delineating the

*Kelleher (1970) describes various systems (time, legislative, financial, technological, operational) which must be coordinated by the decision maker.

consequences of possible solutions in order to decide which solution is the most acceptable. Various motivational and organizational factors create problems in the use of models for this purpose.

Perhaps the specific information needed by the decision maker is not available. For instance, as chemically reactive pollutants are transported in the atmosphere downwind they are transformed into other substances. The models currently available are limited to a range of about 50 km and cannot satisfactorily treat chemical transformation or removal processes.

Another example of this problem can be seen in the area of transportation-control planning. Models are available which indicate the reduction in vehicle miles travelled necessary for a given set of air-quality standards. However, a publicly elected official is not always able to institute programs to reduce car travel locally and may prefer to depend on national emission standards or higher fuel prices. The decision maker only has control over certain variables. As an administrator for an environmental agency and not a transportation agency, he or she cannot design highway projects to reduce vehicle-miles travelled. Thus, what the modeler thinks is important may not be what the decision maker considers important and, as a result, the models are not relevant to the decision maker's needs.

This situation is exacerbated by institutional and organizational arrangements. The model developer may work in a different institution than the decision maker, for instance a consulting firm or a university. As such, it may be logistically difficult to arrange meetings to discuss the work. Also, the objectives of their institutions may differ. The university's department or particular discipline may have more influence in shaping the modeler's research agenda than the government agency. Research done in-house may be more directly responsive to the policy maker's problem.

Another aspect of the institutional structure is the source of funding. If the research is funded at the federal level, even if it is carried out in-house, the needs at the state or local level may not be clearly understood by the federal employees funding the research. If the states are not able to make their needs clear and not able to fund the research themselves, then the necessary technical information may not be available to them.

Organizational arrangements can also aggravate the lack of common interests between researcher and policy maker. The policy problem may require a combination of information across disciplines. If the organizational structure is such that researchers in each separate discipline report to separate administrators, then it may be necessary to create a team which crosses organizational lines. Issues of budget and personnel-hours devoted to the team as opposed to the original discipline-oriented jobs must be managed and a clear definition of the results required must be spelled out.

In other situation, appropriate models may be available but the decision maker does not make use of them. This case can be divided into several problems which arise in the course of the decision-making process. The model results may not be available on a timely basis, the input necessary to run the model may not be available, or the user may not know how to use the model's conclusions.

A typical difficulty faced by a policy maker is that an issue arises in a political context and is perceived as a crisis. Fish in mountain lakes have been dying; the newspapers call it acid rain. Relatively little information may be fueling the crisis and the cause may not really be known. The decision maker is faced with an aroused public and time is not sufficient to design a survey, gather data, and perform a careful analysis*. Instead the decision maker may engage in what has been termed "satisficing". He or she searches for solutions and chooses the first that seems to be satisfactory or acceptable. It may not be the best choice, but it serves the purpose of defusing the crisis**.

In other cases the research may be prolonged as the modeler strives for accuracy by refining the research methodology. However, the policy maker, unable to directly control many variables of the process, may need perspective more than complete accuracy of the numbers. On the other hand, the policy maker responding to crises may find it difficult to engage in long-range planning which would have enabled the research to have started earlier, allowing it to provide results on a timely basis.

Again institutional and organizational arrangements can cause further delay. If research is not done in-house, procedures to procure and monitor grants or contracts can be time consuming. If separate parts of the problem are done by different organizational entities (e.g. modeling by one laboratory and monitoring to collect data by another laboratory), delays can result through the need to communicate up through the hierarchy and back down again.

Even if the research itself is relevant, a frequent problem is the lack of input data required by the methodology or model in order to apply it to the particular situation. Large, complicated models often require detailed data for each application which is only available on a case-by-case basis. Pack and Pack (1977) and the Environmental Protection Agency (1974) show that this is especially true for environmental models. This can be expensive and beyond the resources of the potential user. It might also be the case that the model requires the data to be in a different form from that in which it was originally collected and the resources available are not sufficient to change the form of the data. For example, Brewer (1978) discusses how this is a problem for demographic data.

This mismatch of input data can arise from the dissimilarities in interests and goals of the researcher and the policy maker. The researcher may be primarily concerned with furthering the field of research being studied. The questions of what type of data are involved, what form they are in, or whether they are the minimum amount of data that one can get away with and still solve the problem may be secondary to the questions of whether the theory is sound and the research as accurate as possible. The policy maker with limited resources may be willing to trade off accuracy for the opportunity to use the research.

An organizational structure separating the research and data collecting efforts may create a communications problem which would tend to increase the need for advanced planning. Within the federal Environmental Protection Agency (EPA) the routine data gathering functions are carried out by an organizational group separate from that developing the models. Indeed, much meteorological data is compiled by other federal agencies. Special efforts at communication must be initiated to either develop air-quality models that conform to the available data or, conversely, to develop a suitable data base for the model.

In some situations research may be available which could provide relevant information at reasonable cost, but the policy maker still does not use the resulting data. The

*Downs (1972) describes a cycle of how crises come and go.

**Lindblom (1959) began this opposition to the view of the rational decision maker. Others in this tradition in the policy arena include Shultze (1968) and Moynihan (1969).

policy maker's education and experience may not have included the form and techniques of scientific research. For example, the policy maker's background may be such that he or she is not experienced in using quantitative information and the model results may not be summarized in a readily understandable fashion. Ackerman et al. (1975) show how this led to a politically-based decision even after a large modeling effort by the Delaware River Basin Commission.

The modeler and the user often come from different academic backgrounds and each has his own vocabulary, which can further complicate the process, for example, in modeling the current form of the ambient air-quality standards. The standards are written into law as concentration levels not to be exceeded more than once a year. Using past monitoring records this is a simple determination and in the interests of protecting human health seems suitably cautious. However, the many uncertainties in modeling make such predictions unreliable. From a modeler's point of view a probabilistic approach would be more appropriate, but such an approach is not easily understood by a non-statistician.

Another concern for the decision maker may be the lack of documentation on how to use the methodology. Fromm et al. (1974) assessed the extent of this problem for the US government. Even if the policy maker is experienced in using quantitative information, without documentation detailing the assumptions and limitations of the model he or she may find it hard to determine whether the particular model is applicable. Thus, the policy maker could use the model in the wrong situation, or neglect to use an appropriate model. The lack of validation has similar causes and effects. Sometimes a model is not validated because it is a costly and time-consuming process. Also the modeler may be more interested in starting a new study on a different aspect than in carrying out numerous applications of one model. This is especially true for models requiring large amounts of input data.

On the other hand, when the modeler does explain in detail the assumptions and limitations of the model, the report of the model results may be rewritten and summarized a number of times before it finally reaches the decision maker who is organizationally removed from the modeler. The final report may alter the qualifications or omit them altogether, so that the decision maker is unaware of the limitations and may use the results incorrectly.

2.2 Models Used for Conceptualization

The use of research may not only be a consequence of the policy maker's search for appropriate technical information in a pending decision. A more general use of models is for conceptualization: planning for issues that are likely to arise next or deciding how strategy should be developed to help to deal with issues. Gordon and Gordon (1972) discuss how it may be more difficult to use mathematical models for this type of planning than for problem solving or implementation.

The use of research for conceptualization is not as direct an application as for specific problem solving. However, this type of use may be easier to undertake because the application is less specific and highly accurate predictions may not be required. Thus, the state of the art is not so constraining and relevance is not an overriding concern. General trends may be sufficient, or new input data may not be needed if a previous application was similar to the problems on the planning agenda in the foreseeable future.

The obstacles to research utilization created by the institutional structure may not be as serious in this case. If the need for technical information is not too specific, then research carried out for academic purposes may be useful to the government decision maker concerned with planning strategies on how to analyze a situation.

While availability of information may be less of a problem with conceptualization, the problems of organization and its resulting communications barriers remain. Organizational divisions may hinder the creation of relevant research as well as its communication to the decision maker through the reinforcement of traditional discipline-oriented research and the separation of modeler and user within the hierarchy.

The problem of documentation also arises. If the decision maker is unaware of research results, no matter how applicable, they will not be used.

2.3 Basic Research Leading to Applications

Basic research to expand the state of the art may naturally lead to practical applications and new policies. Examples of this type of use generally come from the physical sciences: biochemical research results in new drug uses, unforeseen at the outset of the research; the transistor leads to portable radios and space satellites.

As a regulatory agency, the US EPA is primarily concerned with operational programs for abating pollution. It focuses on the use of applied research, which in turn depends on basic research. The US Congress is reluctant to fund basic research for a regulatory agency which is responsible primarily for abatement of pollution. Thus, the overall goal of the US EPA would tend to limit the push for basic research. See US Congress (1979) and National Academy of Sciences (1977) for the continuing debate on this subject.

A consequence of this fact is that the basic research to support the applied research may not exist, or may not in practice be applicable to policy issues. This results from the difficulties in communication caused by the divergence of professional interests between researchers and policy makers, as well as their organizational separation, as described above.

2.4 Models Used for Political Advantage or to Advance Self-Interest

Decisions made in a political context require consideration of political and social efforts, as well as technical information. Greenberger et al. (1976) described how these uses can build further support for the wider uses of models for problem solving. The tendency for the modeler, following the scientific tradition, is to strive for accuracy and, failing that, to present research results with appropriate qualifications and caveats. This reluctance to reach firm conclusions for policy purposes leaves that task up to the policy maker. This may enable a policy maker to use research in justifying a decision made for political reasons if the research does not specify that it is not applicable to the situation, and the lack of documentation can reinforce this tendency.

Uncertainty in the scientific data base can also be exploited by advocates in the political process through the institutional structure. The decisions of the environmental agencies in the US are subject to a review process involving the public, whereby evidence, including technical information, is presented by persons representing various interest groups. The lack of conclusive evidence allows these groups to be selective in the evidence they present. If access to this review process were to be institutionally determined and, for instance industrial lobbyists with more resources had greater access, then the information available to the decision maker would also be selective.

Organizational arrangements could also result in the use of models not strictly relevant to the decision. If the modeler is organizationally separated from the decision maker, the usual practice is to communicate the results of the models through a written summary. At each level of the hierarchy it may be necessary to summarize further. In the process the qualifications may be disregarded and partial findings taken uncritically at face value or misinterpreted.

3 CASE STUDIES

In the general description of the possible uses a decision maker might have for models it has been shown that various motivational and organizational factors can influence how models are used in any given situation. In order to investigate how these factors currently affect the use of air-quality models by decision makers in the governmental agencies responsible for environmental affairs in the United States, four case studies* have been conducted.

Two cases involve the permitting of new stationary emissions sources (one an oil refinery, the other a cogeneration power plant). One case involves the effort by a state agency to prepare an implementation plan for meeting the federally mandated ambient air-quality standard for ozone. The fourth case involves the process by which the federal Environmental Agency recently revised that ozone standard.

The 1971 Clean Air Act provides for the establishment of ambient air-quality standards. These are national standards to be attained in all areas of the country. The Act requires that the standards be set so as to protect public health. After standards have been determined by the federal EPA, each state must draw up a State Implementation Plan describing how these ambient standards will be met. For stationary sources, the mechanism used is the issue or refusal of a permit. The state determines for each stationary source how much emission will be allowed so that the ambient standards will not be violated. For mobile sources like automobiles the Clean Air Act has mandated national emissions standards. Thus the states' only methods of reducing pollution caused by mobile sources are either transportation-control planning to reduce the number of cars or attempts to change how and where they are operated.

The first two case studies describe the process whereby permits are issued for stationary sources. The third case is an example of the development of a state implementation plan covering both stationary and mobile sources. The fourth case involves setting one of the ambient air-quality standards. In each case the air-quality models are used to show how source emissions are related to ambient air quality.

---

*The research has not been completed. It is expected to include ten case studies in all. Thus, these findings are preliminary ones only.

3.1 Pittston Oil Refinery

The first case involves the application by the Pittston Company for permission to build an oil refinery in Eastport, Maine. Eastport is a rural community on the northeast coast of Maine near the Canadian Border. It is near a national wildlife refuge and an international park, so that the application is subject to the Prevention of Significant Deterioration (PSD) provisions of the Clean Air Act amendments passed in 1977. These provisions establish increments in air quality which cannot be exceeded by any combination of new projects in such areas. The increment under contention in this case was the 24-h sulfur-dioxide standard of 5 $\mu g/m^3$.

The research and development office of the federal EPA has made several air-quality models available to the states and to individual applicants by means of a computer program. The initial step in the permitting process is the production of an Environmental Impact Statement (EIS). Usually the applicant is required to do the modeling necessary to show what the impact on air quality will be. The Pittston Company used one of the EPA's models to show that the oil refinery's emissions would cause increased 24-h SO_2 concentrations at the park of 4.3 $\mu g/m^3$, which is within the incremental limit of 5.0 $\mu g/m^3$.

The organizational unit within EPA actually responsible for evaluating the application and issuing or denying the permit was its regional office located in Boston, Massachusetts. The regional office conducted its own modeling analysis using different models and came up with figures two to three times larger than the applicant's; the figures produced by the regional office would, therefore, be in violation of the standard. During the next few months, negotiations were held with the applicant. The assumptions concerning proper methods for deriving 24-h concentration levels and what constituted reasonable worst-case meteorology were discussed.

In addition, major design changes were incorporated in an effort to reduce the impact to within the allowable limits. For example, an anchorage site for tanker traffic was eliminated and the sulfur content of the fuel oils was modified. These examples show that models can be used as a design tool to trade off economic efficiency against environmental impact. However, certain emission parameters were also changed, including the number of stacks and the exit velocity from the stack. In this case the project was designed around the model. Without changing the total amount of the pollutants, the model's naiveté was exploited. The models determine ambient pollutant concentrations by relating them to emission rates and not to total emissions. The question of build-up of airborne pollutants in other media, such as water or land, or long-term effects cannot be addressed by the models. Thus, both the applicant and the decision maker ignored these questions because they did not have the means to address them.

Two problems of significance to this study arose during the review of the permit application. The first was that the input data used for the analysis came from a site over 100 km away from the proposed site. Later, data from a site about 25 km away were also used. The unique meteorology of the coastal site for the oil refinery generated much debate on just how representative these data were. Although, under the 1977 amendments to the legislation, EPA can require an applicant to collect meteorological data, no attempt was made to require on-site monitoring in this case.

The second problem concerned the results of a validation study for the model used in this analysis. The one coastal site studied showed that the model underpredicted, which was pointed out in the public comments to EPA regional office. EPA's response was that underprediction had not been found in other studies. However, this evaded the question because other validation studies of coastal sites had not been carried out.

The sequence of events which led up to the final approval of the permit implies another use of the models. The decision maker, in this case the EPA regional office, was also a member of a governmental task force assembled to coordinate the task of siting oil refineries in New England. Thus, one objective of the EPA regional office was to help in finding a suitable site for an oil refinery. Other sites besides the proposed site in Eastport, Maine were discussed in the environmental impact statement, but were rejected on economic grounds. Negotiations were than conducted between the agency and the applicant in an effort to design the project so that it would conform to the environmental constraints. Thus, it was politically advantageous and in the interest of EPA to approve this permit. The way was made easier because the model results were presented as single point estimates to be matched against a single standard value. Worst-case meteorology was used and no sensitivity analysis was done to account for the possibility of underprediction. Moreover, the documentation and guidelines for choosing assumptions were ambiguous. The EPA regional office was able to pick one end of the suggested range for the wind speed, for example, without further justification other than the statement that "it appeared in the guidelines".

It was the job of the scientists in the EPA regional office to run the models, justify the assumptions, and respond to public comments. Discussion with the regional administrator, the ultimate decision maker, seems to have been limited to the legal question of when, not whether, the permit should be issued. Questions of how accurate the model results were or what environmental impacts may not have been measured were not apparently discussed. Although the allowable increments at issue here appear to be of the same type as the other standards that EPA administers, they are easier to violate and demand more precise measurement. Because the decision maker was not familiar enough with the air-quality models to make this distinction, he had to rely on the analysts. Yet the analysts did not raise this issue because they did not have other methods to draw on.

To summarize our findings in this case study, the models were used appropriately to redesign the project to reduce the impacts on air quality. They may have been used inappropriately if long-term and other impacts are considered, because the models are not equipped to deal with such problems. The analysis also may have been used to justify a decision taken on other grounds. This was facilitated by the lack of familiarity of the decision maker with the models, the tendency within EPA to emphasize worst-case analysis, and the lack of validation studies and documentation. The end result of these tendencies is one point estimate compared with the standard level, without reference to accuracy or other uncertainties.

3.2 Harvard Power Plant

The second case study also involves issuing a permit for a stationary emissions source, this time in an urban rather than a rural area. The source is a power plant using diesel engines to generate both steam heat and electricity, and the controversial pollutants in this case are nitrogen oxides. The applicant, Harvard University, proposed building the plant near downtown Boston, and produced an Environmental Impact Report (EIR) which contained modeling results. The applicable standards were easily met once the project was redesigned to exclude an incineration facility. The state of Massachusetts' environmental agency is the designated decision maker in this case. At this stage in the process comments were received from the public on the application and draft EIR; these comments criticized the assumptions and the input data used for the model analysis. The input data on background levels were from a site close to the plant site, but it appeared to be a "hot spot", unrepresentative of the area because of the high density of traffic. The applicant began monitoring for background levels on nearby but more-representative sites in an effort to justify their selected input parameters.

Other problems were not dealt with so easily. Public concern soon centered on the short-term NO_2 levels which would be generated by the operation of the diesel engines. Because the federal EPA had not set a short-term standard for NO_2, the model results could not be compared to a standard. Instead the citizens' groups opposing the construction of the plant focused on the evidence of adverse health effects caused by NO_2. The state agency had always based its decisions on an already-established national standard and were unprepared to deal with the lack of a standard. Normally, the federal EPA was the agency which would have dealt with the health-effects evidence.

The state agency had been initially inclined to issue the permit, but the controversy over health effects prompted it to solicit more information. It sought the opinion of health experts on what concentration levels cause health effects, and required the applicant to do more detailed modeling in an effort to refine the estimates of the emissions impact.

The health experts' opinions varied considerably. The applicants found experts who proposed a high concentration level, while the citizens' groups found other experts who suggested caution and a low level. The new model analysis was more complex but, when the modelers were asked what the level of accuracy was, they could say only that it was within a factor of two. The result was that, because the 1-h health-effects threshold and the ambient air-quality model results were both estimated to be within the same range, the policy maker did not have a clear picture as to which way the decision should go. This case, as opposed to the Pittston case, involved much more consideration of what the models did and did not indicate. The lack of a standard seems to have pushed the decision makers into examining more closely the accuracy and uncertainties of the model results. However, another factor was instrumental in keeping the controversy alive. All the participants in this decision were geographically situated in the same area and access to information was relatively easy. In the Pittston case the citizens' groups opposing the permit were located in northern Maine and in Washington, D.C., while the decision maker and applicant were negotiating in Boston. This caused delays in obtaining information so that the public comments were often no longer relevant by the time they reached Boston. In both cases public comments seemed to have been effective in raising questions about aspects of the model assumptions which might otherwise have gone unnoticed by the government modelers who were reviewing the applicant's analysis.

The role of the citizens' groups or public-interest lobbies points out another use that was made of models in the Harvard case. Several citizens have stated that they are opposed to the construction of any power plant in an urban area and that they specifically opposed the expansion of a large institution like Harvard University into their residential neighborhood. Thus, they present all information that can be used against the applicant, and only such information. Of course, the applicant also presents only such information as is advantageous to his position. But that is the purpose of the public hearings, to hear both sides of the issue.

For the decision maker who is not familiar with the details of the modeling, however, it becomes a question of how to choose among the experts. Numerous lengthy hearings have been held on the matter to ascertain how the experts arrived at their conclusions. The decision has shifted several times over this period as new issues are raised and the final approval or denial of the permit has, at the time of writing, not yet been promulgated.

To summarize, the models in this case were used to redesign the project to meet applicable standards. However, the lack of a short-term NO_2 standard meant that the decision maker and applicant had to redefine the problem to include this issue. Public interest groups fueled the controversy for several years and pushed the decision maker into considering just what the various assumptions and the model's methodology and accuracy implied.

3.3 Connecticut's State Implementation Plan

The third case involves an implementation plan for controlling pollution sources that must be developed by the state of Connecticut so that the national ambient air standards can be met. The particular portion of the plan that this case investigates is the ozone standard.

When work on the implementation plan first began in Connecticut, the federal EPA required the use of a technique called Appendix J, which is a modified form of rollback. The graph to be used by the states is based on data from such cities as Los Angeles; therefore Appendix J was frequently criticized as being inapplicable to other cities not included in the graph. The method for generating the graph which would be specific to a particular city required five years of data on ambient pollutant concentrations. This type of data was available in most cases. Thus, many states put pressure on the federal EPA to allow them to use the linear rollback method. The only other types of alternative were the complex photochemical-dispersion models which required even more input data.

These difficulties with the methodologies available for determining the reduction in the precursors to ozone led to an effort by the federal EPA to develop another model. The objective of this research effort was to develop a technique for modeling the relationship between the precursors to ozone and ozone itself which would take into account nitrogen oxides as well as hydrocarbons, would be based on the physical and chemical nature of the pollutants, and would necessitate only a limited amount of input data. The result of this effort was EKMA, an approach based on smog-chamber data and using isopleths to determine the necessary reductions.

This, however, took time. Meanwhile, the states had also been struggling with the problem. The state of Connecticut had produced its own model consisting of a fixed-grid system with an extensive inventory of source emissions superimposed on it, in conjunction with assumptions about how the sources mixed and were dispersed within each grid.

One of the primary objectives of Connecticut's effort at model development was to take into account the pollutant-transport problem. Connecticut is downwind from the highly industrialized areas of New York and New Jersey. It was maintained that much (perhaps up to 70%) of the problem was due to pollutants emitted in these two states and, hence, was their problem rather than Connecticut's. The model estimated the effects of transport and reduced emission-reductions requirements accordingly.

The state environmental agency encountered opposition from several quarters. The state legislature and transportation agency did not like the fact that the model results showed that large areas of the state would have tight restrictions on what highways and other projects could be built. When concessions were made to the transportation agency and the legislature, environmental lobbying groups concerned with public health and environmental impacts began to oppose the model results. The federal EPA criticized the theory and assumptions of the model. The initial federal approach to the ozone problem had not taken transport into account. Thus, while they were studying the transport issue, the policy had not been changed. In order to meet the national standards, states still had to base their reduction estimates on existing ambient air-quality, regardless of its jurisdictional source. Also, since the job of evaluating state plans would be considerably harder if each state developed its own model, all states were required to use Appendix J.

The primary problem facing the state agency was to persuade the federal EPA to recognize the transport problem. This fact, combined with the opposition within the state to the model's results, led to Connecticut's own model being abandoned.

Several alternatives were acceptable to EPA, thus removing one obstacle. However, both Appendix J and linear rollback were still unacceptable to the state, and the photochemical dispersion models required money and data resources not available. By this time, however, the federal EPA developed the EKMA model. Although the state still did not fully agree with how this model handled the transport issue, the state modelers decided to use the EKMA model because it was a methodology acceptable to the federal EPA.

A cycle in the planning process developed. The state would fund a model-application effort. Both the state transportation agency and the federal EPA would find fault with the model. The transportation agency would suggest another study and, lacking support for the environmental agency's modeling efforts or even under threat of sanctions from the federal EPA, the state legislature would agree to further study. Unless forced to implement a plan, Connecticut saw no reason for doing so. This is the situation as it stands at the time of writing. Most of the old issues remain unresolved while new ones, also not handled well by existing models, continue to arise.

The case points to the problem in communication and divergence of goals that exists in our federal-state system. The problem that the state was attempting to model was different from the problem as defined by the federal EPA. Much time and effort was spent in developing a model specific to the state's definition of the problem. Finally, this was abandoned because it did not meet the criteria of the federal EPA.

The federal EPA, on the other hand, views the process as linear, in one direction. It develops the models, then the states use them. Because it is familiar with the models and all states are treated in the same way, EPA's job is made easier by not having to assess each state's model as well as its own plan. However, this does mean that it resists learning from the states and is slow to perceive and consider new issues.

3.4 Revision of the Ozone Standard

The fourth case focuses on federal rather than state actions. In this case a cost and economic-impact report was prepared by the operating program office of the federal EPA. The report estimated the nationwide costs of achieving alternative levels of an ozone ambient air-quality standard. Among the operating program office's responsibilities is the preparation of guidelines for the use of the air-quality models by the states. Although not normally involved in developing the models, the operating program office worked with the research office to develop the EKMA isopleth model mentioned in the previous case.

The report began with a sample of eight cities. This was later expanded to include large cities and was therefore no longer just a sample for a data base. The EKMA model, which is most suited to high-density, urban areas, was used in the initial drafts of the cost report. However, when the analysis was expanded the EKMA model did not change the input parameter, the HC/NO_x ratio, which makes the model results specific to each city. City-specific ratios were therefore not available. A sensitivity analysis was carried out, however, which demonstrated that the results were not sensitive to this parameter.

The use of the model changed in later versions. First, linear rollback was introduced as another estimate in a range of estimates. This was justified because city-specific data were not available and nationwide reduction estimates do not require a high degree of accuracy. These estimates were considerably lower than those using the EKMA model.

It should be noted that the cost estimates were not required by the Clean Air legislation. EPA has interpreted the act as saying that economic costs of control should not be considered in setting the ambient standards; health effects are the primary concern. The cost estimates were required administratively by the President. EPA aimed for an order-of-magnitude estimate only to satisfy these requirements and to save their resources for the consideration of health effects.

The cost figures, however, were controversial. They ranged up to an annual cost of US $12 billion. Thus, although technical reasons were cited in the report for the choice of models, the persons involved in the writing of the reports also cited the pressure within EPA to ensure that the costs were as low as possible. The initial choice of EKMA seems to have been one of "what model am I familiar with" and "what model uses the latest, new technique", rather than a complete search for what was most appropriate. Indeed, the operations office did not consult the developers of the model in the research office about the choice. As pressure mounted to keep the cost estimates low, linear rollback was introduced. In the final version and in presentations to the EPA Administrator only the linear-rollback figures were cited, because it was felt that listing different ranges and estimates from the various models was too confusing. While the choice of model may have been correct for the problem as defined by EPA, it was simplified by political considerations.

Meanwhile the federal policy on the Appendix J methodology was being revised. As mentioned in the Connecticut case, the states were required to use Appendix J in formulating their state plans for achieving the ozone standard. There had been complaints from other states besides Connecticut and the EKMA model had been developed in response. The change in the standard provided the opportunity to revise the regulations regarding the use of models. The new regulations allowed any of the methods to be used so long as the application of all reasonably-available control technologies was assured. States attempting to show attainment without the application of such control technologies would have to employ photochemical-dispersion modeling. This revision in the regulation was accomplished without controversy. No one objected to replacing a method which had been shown to be in error by the most flexible of approaches.

4 CONCLUSIONS

Four possible types of uses of models were outlined in the introduction: for problem solving, for conceptualization, as a result of basic research, and for political advantage or self-interest. Examples of all of these have been given in the case studies.

A major problem encountered in using the models to issue permits or plans where new emissions sources might be built was the lack of input data. Accurate meteorology and monitoring is essential if the worst case is to be chosen rather than some average. In the Harvard case the applicant supplemented the available data with new monitoring sites because it was advantageous to do so. Since it was neither advantageous nor required in the Pittston case, no new input data were collected and the problem remained unsolved.

The emphasis on conservative, worst-case analysis produces another problem. Since the models are used to produce one point estimate, no sensitivity analysis or allowance for uncertainty in the data or the methodology is presented to the decision maker. This is encouraged by the decision maker, who usually is not a scientist and therefore has difficulty assessing variable results. When only one number, which is below the standard level, is presented, the decision is obvious, and the decision maker does not have to question the accuracy or indeed the applicability of the model. These questions did arise in the Harvard case because no standard had been set. But even here, the short-term NO_x effects were not seen as an issue until they were raised by public comments.

This indicates a major influence on how models are chosen and used in problem solving. The choices made are based on experience, but this experience is not that of validation and documentation since they did not exist to a large extent. Instead the experience is based on what was used last time. For the decision maker this means point estimates of concentration levels have become familiar. Whether the potentially significant deterioration provisions are substantively different and demand a different approach is not asked. If no standard exists, the tendency is to avoid the issue.

For the modeler it is easier to use a model that has been used previously rather than to learn how to apply a new one. Also, if the decision maker is satisfied with point estimates, then the work of producing sensitivity analyses and educating the decision maker as to what they mean is not worth the effort. The state of Connecticut also learned that it is easier to use a model that is familiar to the federal EPA than to fight this issue as well as the transport issue with both EPA and its own state agencies.

Besides the uses of the models in problem solving we have shown several instances of more political uses. The process of public hearings is an advocacy system. Lobbyists for the applicant and for citizens' groups choose their models and information to support their position. Several times key issues regarding the accuracy and appropriateness of the models were initially brought up through the public-hearing process. In a society of limited resources this may be a fairly good way of making sure such issues are raised.

EPA or the state agencies cannot have the resources or the time to investigate every aspect. However, a problem with this approach is that it tends to occur late in the process so that the EPA modelers are put in the position of having to justify their conclusions. This may be exacerbated if the distribution of resources or availability of information is unequal. In the Pittston case the citizen advocates were at a disadvantage on both counts. In the Harvard case the citizen groups had better access to the information.

As for the other two uses – conceptualization and the use of basic research – fewer examples were found. The accumulation of state experience in trying to use Appendix J led to a search for a better method. But policies on transport, on monitoring for permit applications, or on the potentially significant deterioration provision had not yet been developed. The available models are used but there is no strategy for how they should be used.

Basic research to develop better, more accurate models had concentrated on the photochemical-dispersion models. But these require data and computer time beyond the resources of the state agencies, and therefore, are not often used. It was because this research was not responsive to the decision makers' needs for more timely and less expensive models that the EKMA model was developed.

Our initial cases point out one reason why the experience of the states in using the models has had little impact on the broader issues of strategy and policy. The process is, by and large, one-way. The federal EPA has the resources to develop the models, which are then made available to the states. Thus, only seldom does a state search for other options. Moreover, the federal EPA is organizationally structured so that the research office is separate from the office which helps the states in their model applications. This separation of the model development and use activities, both within the federal EPA and between it and the states, hinders communication of the states' experiences. Officially the federal EPA provides technical assistance to the states. Only a few administrators have realized that the states' experiences may also be useful to the federal EPA.

These conclusions are, of course, preliminary. Three more case studies are underway and we expect to complete ten in all. Our aim is to expand the examples to other states and other programs and to continue investigating how the motivational and organizational factors influence how the models are used.

ACKNOWLEDGMENTS

This research was supported by the US Environmental Protection Agency, under Grant R805558-01.

REFERENCES

Ackerman, B., Rose-Ackerman, S., Sawyer, J., Jr., and Henderson, D. (1975). The Uncertain Search for Environmental Quality. The Free Press, New York.

Brewer, G. (1978). Operational social systems modelling: pitfalls and prospectives. Policy Sci., 10: 157–169.

Downs, A. (1972). Up and down with ecology: the issue–attention cycle. Public Interest, 28: 38–50.

Environmental Protection Agency (1974). A Guide to Models in Government Planning and Operations. Environmental Protection Agency, Washington, D.C.

Fromm, G., Hamilton, W., and Hamilton, D. (1974). Federally supported mathematical models: survey and analysis. National Science Foundation, Washington, D.C.

Gordon, M. and Gordon, M. (1972). Environmental Management: Science and Politics. Allyn and Bacon, Boston.

Greenberger, M., Crenson, M., and Crissey, B. (1976). Models in the Policy Process. Russell-Sage Foundation, New York.

Kelleher, G. (Editor) (1970). The Challenge to Systems Analysis: Public Policy and Social Change. John Wiley, New York.

Lindblom, C. (1959). The science of muddling through. Public Admin. Rev., 19: 79–88.

Moynihan, D. (1969). Maximum Feasible Misunderstanding: Community Action in the War against Poverty. The Free Press, New York.

National Academy of Sciences (1977). Perspectives of Technical Information for Environmental Protection, Washington, D.C.

Pack, H. and Pack, J. (1977). Urban land use models. Policy Sci., 8: 79–101.

Schultze, C. (1968). The Policy and Economics of Public Spending. The Brookings Institution, Washington, D.C.

US Congress (1979). Authorization for the Office of Research and Development. Environmental Protection Agency, Washington, D.C.

Weiss, C. (1978). Improving the linkage between social research and public policy. In L.E. Lynn, Jr. (Editor) Knowledge and Policy: the Uncertain Connection. National Academy of Science, Washington, D.C.

REGIONAL AIR-QUALITY POLICY ANALYSIS

R.L. Dennis
Environmental and Societal Impacts Group, National Center for Atmospheric Research, Boulder, Colorado (USA)

1 INTRODUCTION

Air-quality analysis depends heavily on computer models to predict the future effects of present actions. Imperfect as these models may be for the questions that are asked of them, they are still the best means of obtaining the needed predictions.

There are three main factors which sometimes make the use of computer models for environmental management (including air-quality analysis) a difficult task. First, many questions that are asked of air-quality models are at the limit of our present scientific knowledge and expertise. Thus the predictions of the models can have a large degree of scientific uncertainty which is difficult to incorporate into a decision-making process. Second, modelers tend to go into exhaustive detail in the models, attempting to include all the known scientific information. The profusion of detail does not necessarily make the model more useful for decision making. On the contrary, critical parameters may not be clear to the user, who may not be able to distinguish easily critical responses to policy actions in a pattern-recognition mode of first-order analysis. Third, the assessment of the technical predictions of a model is complicated, and may be distorted, by the fact that the technical information is only one of the many factors that enter into policy decision making.

Methods need to be developed to improve the use of predictions from models or from technical assessments for decision making related to environmental management, particularly air-quality management. This paper is a report on research in progress to this end at the National Center for Atmospheric Research and the University of Colorado at Boulder. There are two major thrusts to the research.

(a) To further develop simplified predictor models of ensemble averages relevant to policy analysis. Such models emphasize the pattern recognition of critical features of a system and allow the development of scenario models for a great number of rapid adaptive assessments.
(b) To develop further a method for linking technical analysis with the analysis of social values. Such a linkage may provide a more appropriate means of asking the right questions and interpreting technical information.

These methodological developments are being pursued in the context of a case study – the Denver Regional Air-Quality Management Study. In this paper we give a general overview of this study, highlighting the air-quality problem of Denver, Colorado. We list a set of models chosen to define a fast scenario model for the policy analysis as well as the scenario model itself. The social-value model and the linkage system are qualitatively described. The use of these methods for air-quality policy analysis is then discussed by presenting the results of a two-day workshop held for a number of planners, policy makers, and concerned individuals in September 1979. We conclude with a brief discussion of some of the major conclusions to be drawn from the experience of that workshop.

1.1 The Problem Setting

The air-quality problems of the Denver metropolitan region may be characterized from two perspectives.

(i) Violations of federal air-quality standards: CO violations in the winter; oxidant violations in the summer.
(ii) Public concern about air quality: concern about human-health effects; concern about visibility degradation and pollution coloration (the "brown cloud").

Several aspects of the problem setting should be noted.

(1) The public assumes a close connection between health effects and visibility degradation. This assumption is fallacious because the pollutants causing potential health effects are different from the pollutants causing potential visibility degradation. Most importantly, different sources contribute differently to these two classes of pollutants. A program may significantly reduce health impacts yet be judged a failure by the public because visibility degradation has become worse.
(2) Concentration on the pollutants regulated by federal legislation has tended to mask the visibility issue in the past. Hence insufficient data are available for a complete characterization of visibility degradation.
(3) Institutional divisions, the mandate of the Clean Air Act Amendments of 1977, and lack of manpower have forced attention almost exclusively on mobile sources alone instead of on all sources simultaneously and in equal detail.
(4) There are some foreseeable future impacts, e.g. severe visibility degradation due to diesel-car emissions, for which decisions need to be made now without waiting for the impacts to occur. Diesel cars look attractive from the point of view of energy efficiency and are being sold in increasing numbers.

Because sources contribute differently to different components of the air-quality problem in Denver it is important to develop a comprehensive regional assessment capability that includes both stationary and mobile sources. Such an assessment capability must address the entire causal chain. Starting with economic activity and population growth, it must derive the emissions resulting from these activities, determine the relevant pollutant exposures resulting from the emissions (through dispersion, air chemistry, and transport), and finally project the effects of exposures to these pollutants:

Activities → Emissions → Exposure → Effects

1.2 Formulation of the Analysis

The scenario model for the causal chain given above must be capable of addressing numerous options easily. It must be capable of providing sufficient detail for important aspects of air quality or important policy options, yet be adaptable to changing questions (as some options are eliminated, new ones will arise). The information needs of the process of policy analysis are not static.

These requirements are achieved by building the scenario model in component form, where many of the components consist of an appropriately aggregated and simplified version of a more complex and complete model. The simplification allows for greater computational speed when using the component in the scenario model. More importantly, however, the building of the simplified model from the more complex model forces a more complete examination of the underlying pattern of response of the complex model. This understanding is distilled into the simplified model, giving a much greater clarity of understanding of the system being modeled (or of the predictions of the complex model that is supposed to represent the system). By using components the scenario model can be easily adapted to different questions (through the addition or further development of components) without the necessity of changing the entire scenario model. The building of these simplified predictor models is the first thrust of the research being conducted in the context of the Denver case study.

The linking of the scenario model to models of social values is the second thrust of the research. For problems such as the management of the air quality in Denver it is important to make the social values involved as explicit as possible. There are several reasons to go further and to quantify and computerize the social values for explicit quantitative linkage to the scenario model.

(1) Problems as complex as the air quality in Denver produce a tremendous cognitive overload. Quantification of the social values provides a "bookkeeping" system for what is happening in the analyses and also provides a consistent and reproducible means of analyzing hundreds of scenarios.
(2) The ability to operate the analysis with the social values explicitly linked to the technical assessment may circumvent role reversals between fact and value and eliminate false arguments as to what are "facts".
(3) The use of social values in this quantified and precisely defined manner is helpful for the sifting and evaluation of hundreds of scenarios or option combinations and may be a good way to improve communication of the results of policy analyses.

1.3 The Research Process

It is essential that the potential user should understand the process of adaptive-scenario model building and its linkage to social values. This is particularly true for large multidisciplinary problems such as the pollution in Denver. Otherwise, the methodologies and models that are developed will not be well communicated or well received, and hence they will not be used. In the Denver study the approach that was chosen to bring the potential user into this process was that of working sessions and update sessions with the most active potential users coupled with a periodic workshop for a broader set of potential users (planners, policy makers, and interested citizens).

The working sessions help to define issues in detail and to pinpoint components of the scenario model that should be developed during a particular phase of work. The workshop, in contrast, presents to all groups the full picture of the analysis to date. A sense of the current state of the process is communicated. An overview can be seen by all and feedback on the directions of the research can be obtained from the potential users while there is still time for the scientists to respond. The adaptive process is in operation, even while new methods are being developed to ensure the relevance of the scientific work. For the next phase of work the process is repeated, with a return to work sessions. By thus involving the potential users in the research process the researchers know better whether they are responding to the real decision-making needs of those who shape policy.

2 THE SCENARIO MODEL

2.1 The Underlying Models

An entire set of models, shown in Table 1, is necessary to develop adequately the causal chain of activities → emissions → exposure → effects.

2.1.1 Activities

An input–output model that was developed for the State of Colorado is used to provide a consistent and sectorially-disaggregated projection of economic growth and the concurrent growth in households needed to satisfy labor-force requirements. This model provides (1) sufficient disaggregation (ten sectors) for the projection of the relevant stationary-source emissions and (2) consistent economic and household projections necessary for the prediction of transportation needs for the mobile-source emissions model.

2.1.2 Emissions

The stationary-source emissions model is a simple accounting model. Matrices of fuel use per unit of economic activity and per household convert the output from the input–output model to fuel demand, with allowance for fuel substitution in the future. A second matrix of emission factors per fuel type calculates the emissions according to stationary-source type (e.g. residential and commercial space heating or emissions by industrial processes).

Mobile-source emissions are calculated by a model with more complex choices because so many of the policy options interact at various stages in the mobile-source link between activity and emissions. The act of trip taking, the choice of trip mode, choices within a mode, and technological emission-control options are all modeled to provide the necessary decision–response points for the policy choices concerning mobile sources.

TABLE 1 The underlying models.

Activities model	*Pollutant concentration models*
Input–output model of population and economic activity	CO-concentration model
	Haze model
Emissions models	Oxidant-concentration model
Stationary-source emissions model	
Mobile-source emissions model	*Effects models*
Trip-generation model	Human-health impact model
Vehicle-miles traveled model	Standards-violations model
Vehicular emission per mile model	Aesthetic-impact model
	Energy-use model

2.1.3 Exposure or Concentration

Three different simplified predictor models are being developed for CO, oxidants, and haze (visibility degradation). The models will be based on detailed three-dimensional dispersion models, detailed photochemical and aerosol-chemistry models, and detailed three-dimensional dispersion models with simplified chemistry. At the September 1979 workshop only the CO predictor model was available for use in the scenario model.

2.1.4 Effects

Four effects models were included in the scenario model at the September 1979 workshop.

(1) CO health effects. The indicator used was the relative change from present production of severe aggravation due to CO exposure in members of society with severe heart conditions. This relative change was rather insensitive to the starting point on the logit dose–response function of the California Air Resources Board which might be assumed to represent the present level of aggravation.
(2) Haze. Visibility degradation was defined in terms of the measure of haziness produced by fine particulates ($< 1\ \mu m$) based on Mie theory. The relative contributions to the scattering by particles from industry, space heating, gasoline vehicles, and diesel vehicles were accounted for in this measure. As with health effects, changes relative to the present were modeled.
(3) Energy use (mobile sources). This model was based on the US projections of fleet fuel economies as presently mandated. Again, it was considered that relative change from the present level was the most appropriate indicator to project.
(4) Standards violations. The expected number of 8-h CO-standards violations at the monitoring station in central Denver for the winter season was projected. This assumed no changes in the meteorological constellation from year to year, which is a simplification. The projected number is therefore only a measure of an expected value around which there is a band of uncertainty.

2.2 The Scenario Model

The scenario model was composed from the above set of models, simplified predictor models, and direct-model components. It is schematically shown in Figure 1 together with

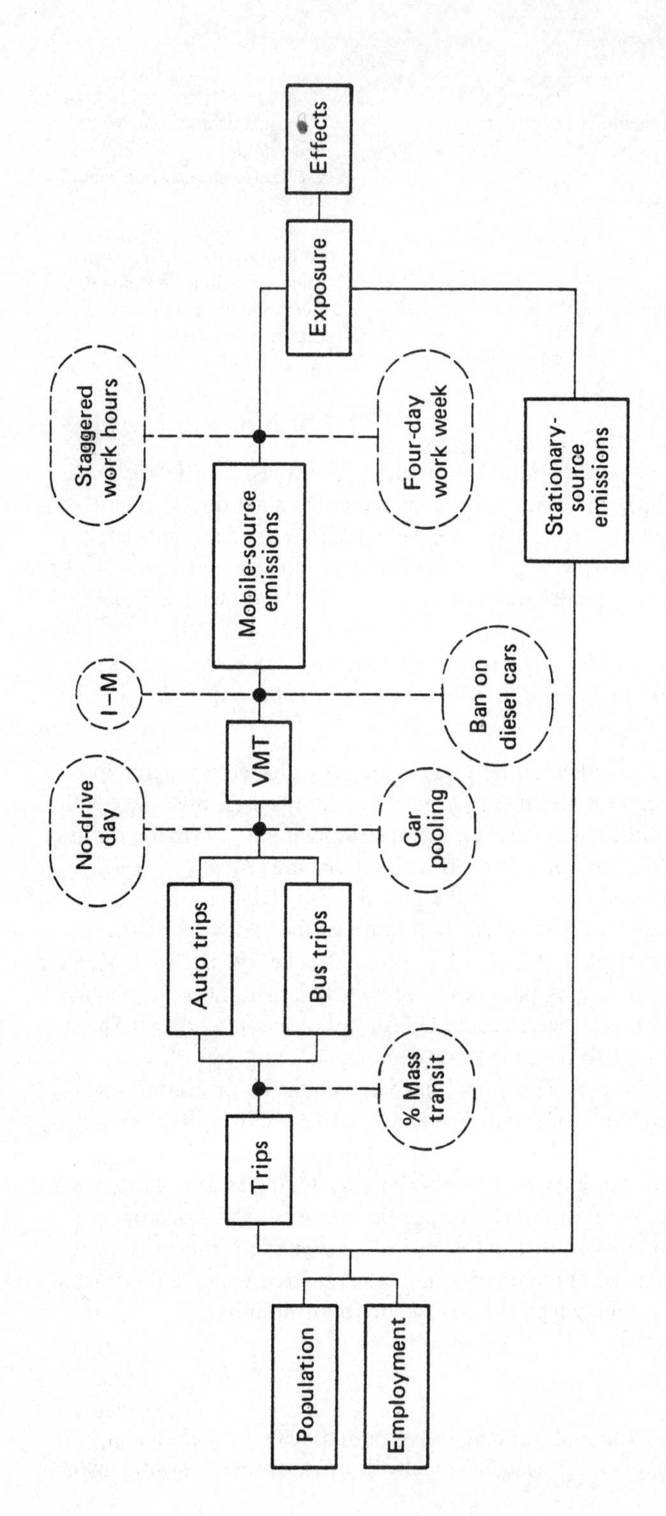

FIGURE 1 A schematic diagram of the scenario model with the policy options that affect it.

the policy options that were considered during the September 1979 workshop. The diagram shows the points in the causal chain of activities-to-effects that the policy options actually affect.

The policy options are all concerned with mobile sources of pollutant emissions because of the interest in these options in the Denver region. The policy options were activated in a binary-switch mode – they were either ON or OFF. Reasonably extreme values of possible effects were taken to assess the full potential for possible change. The mass-transit option assumed a tripling of its present share of the model split by the year 2000, more than tripling the numbers using the mass-transit system. A larger increase would be unrealistic because of capital constraints.

The car-pooling option assumed an increase in car pooling for work trips from the present 20% to 50%. The no-drive day option assumed an across-the-board reduction of 10% in nonwork trips. The inspection–maintenance option assumed optimistic reductions in the rate of CO emissions for cars one year old and older. The diesel-car ban would ban new diesel cars from being sold in Colorado after 1981. In the model it was assumed that 25% of the new cars would be diesel by the year 2000 (a percentage that now seems to be too low, given more-recent predictions). This option was introduced because of the deleterious effect diesel-car emissions have on visibility. At the same time, however, the use of diesel cars is being advocated because of their energy efficiency. The four-day work week option does reduce the number of vehicle-miles traveled by reducing the number of drives to work. The staggered work-hours option is aimed at changing the diurnal pattern of emissions by spreading the diurnal peaks and thus reducing the maximum hourly emission (tons per hour) while not affecting the total daily emission. This is also the primary aim of the four-day work week. These last two options were assumed to be mutually exclusive. Both options assumed that 30% of the work force was participating in the program.

The binary-switch method of analyzing the options was chosen for clarity of presentation and analysis. With two mutually exclusive options there were 96 possible combinations to be analyzed for a given scenario of economic and population growth. To examine the effect that uncertainty in economic and population growth would have on the overall desirability ratings of a set of policy decisions and to begin examining means for addressing uncertainty, three scenarios of economic and population growth spanning the range of projections made by various agencies were defined. The middle scenario was loosely termed the base case. Thus a total of 288 scenarios were run with the scenario model and put into a data bank for linkage with the social values to be derived during the workshop.

The base-case (do-nothing) scenario is shown in Table 2. The health effects due to CO drop very rapidly to very low levels relative to current levels and remain low until the year 2000. After 2000 they begin to rise fairly rapidly to 28% of current levels by 2020. The effectiveness of technological emission control is being overtaken by population growth and the resultant growth in Vehicle-Miles Traveled (VMT) (Figure 2, compare the VMT and CO curves). Haze, in contrast, never really improves, becoming more than three times worse than at present by 2020. Part of this deterioration is due to the entry of diesel cars into the car fleet, but the emission of small particulates ($< 1\ \mu m$) also increases steadily (Figure 2, the PM curve). The emissions of small particulates are fairly evenly divided between mobile sources and stationary sources, even though the latter use predominantly natural gas. This implies that any fuel switching for the stationary sources could contribute

TABLE 2 The base case of the scenario model and the policy-option switches.

Year	Outcome				
	Health effects[a]	Haze[a]	Standards violations[b]	Energy use[a]	Side effects[c]
1975	100	100	40	100	0
1980	22	99	21	98	0
1983	3	100	8	92	0
1985	0	103	3	87	0
1987	0	109	1	84	0
1990	0	122	0	81	0
1995	0	147	0	81	0
2000	0	175	1	83	0
2010	2	246	7	105	0
2020	28	338	23	145	0

Intervention pattern

Car pooling	OFF	Diesel-car ban	OFF
No-drive day	OFF	Staggered work hours	OFF
Mass transit	OFF	4-day work week	OFF
Inspection–maintenance	OFF		

[a]Percentage of current level.
[b]Per season.
[c]Out of 10.

significantly to further deterioration of the visual air quality and also that if only mobile sources are considered particulate emissions may not be reduced sufficiently even to allow Denver to maintain its present level of visibility degradation, a level that is considered poor by the Denver community.

CO-standards violations in the winter season decrease (as do the health effects); they do not decrease as rapidly as the health effects, but still become acceptable by 1990. They then begin to increase and will become a problem again after the year 2000 (note the curve for CO emissions in Figure 2). Diesel cars actually help with regard to standards violations, although they worsen visibility. Without the entry of diesel cars into the market, the CO-standards violations as projected in this scenario model would never reach acceptable levels. Energy use by vehicles is expected to decline below present levels but then to increase beyond the present levels by the year 2000 if no further attempt is made to improve fleet-year fuel efficiency beyond the 1987 targets. This projection of energy use assumes no radical change in driving behavior due to increased fuel prices.

The next output, side effects, is very important and will be discussed in more detail in Section 3. This variable effectively characterizes the poorly quantifiable side effects (effects not directly considered in the scenario model) that might occur when each policy option is turned on. Since no options are turned on in the base case (the do-nothing scenario) the side-effects variable is zero.

3 THE SOCIAL-VALUE MODEL

As previously stated, we wish to evaluate the scenarios via social-value models through a linked computer system rather than to try to proceed with a less well-defined

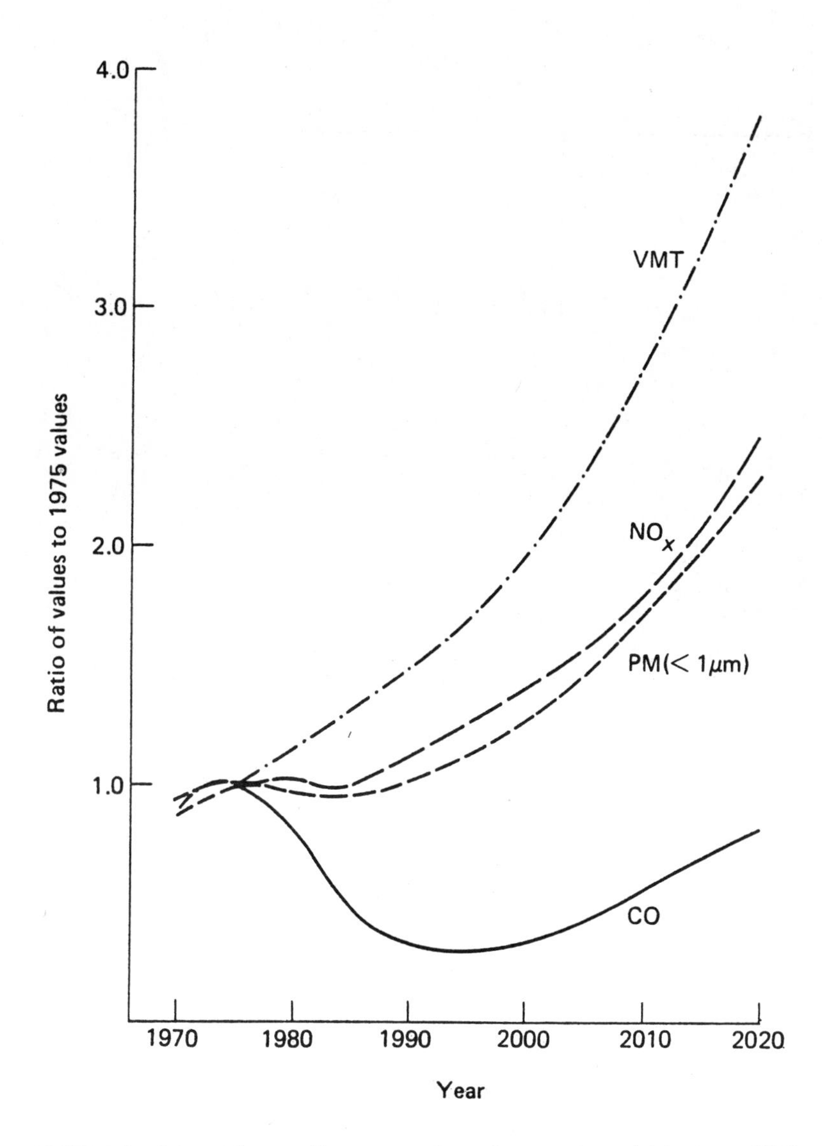

FIGURE 2 Selected intermediate outputs from the scenario model (ratios to 1975 values).

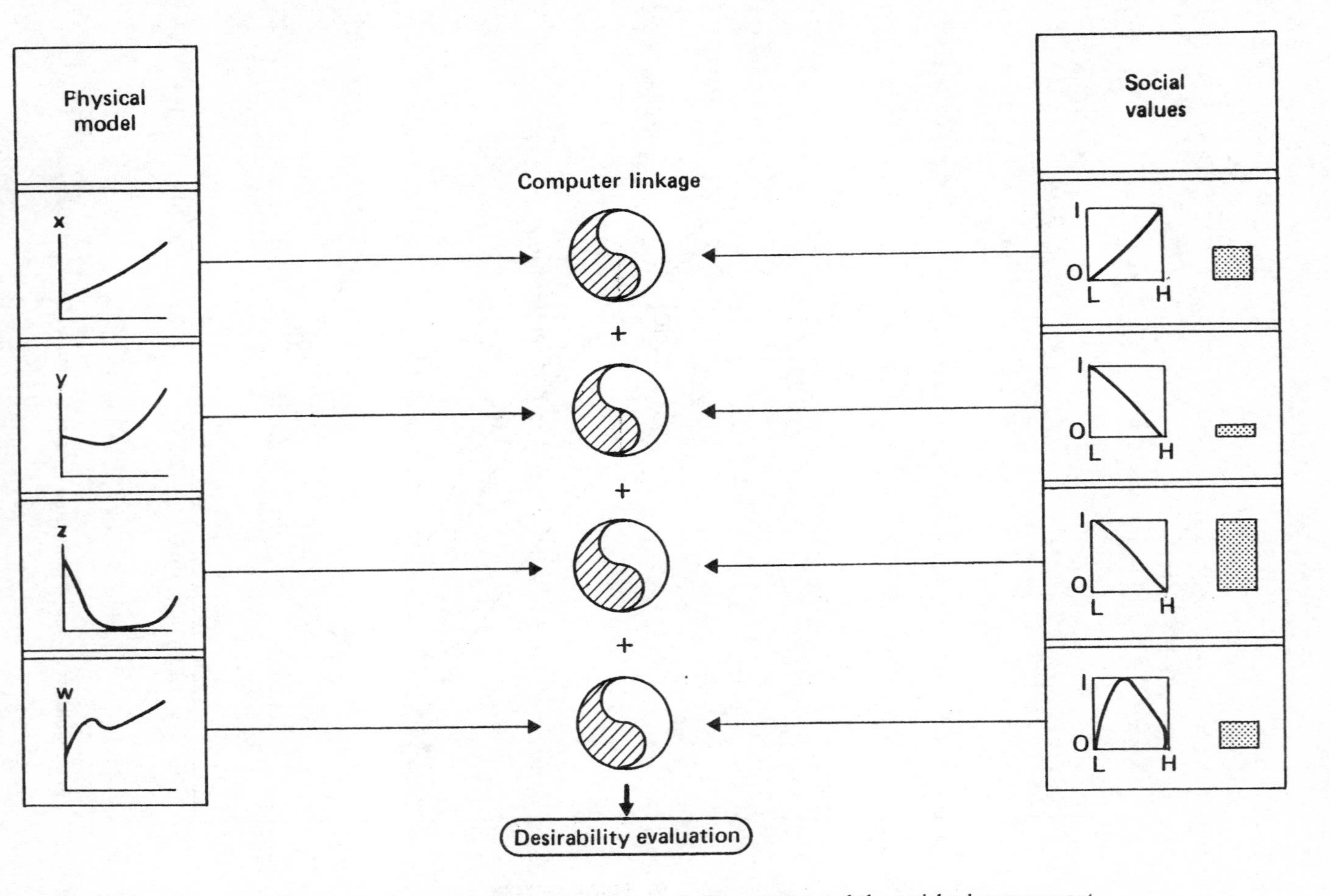

FIGURE 3 A schematic diagram of the symmetrical linkage of the technical assessment and the social-value assessment.

discursive analysis of the scenario results. The aim is to provide a more quantitative assessment as well as, hopefully, a more understandable package for communication purposes. This linked system is shown schematically in Figure 3.

The social values are expressed quantitatively in terms of function forms (the graphs) and weights (the bar associated with each graph). The function form is the functional relationship between the numerical values of the variables and the individual's judgments of the desirability of these numerical values. The numerical values start with the lowest magnitude, which occurs in the scenario set at the origin, and end with the largest magnitude, which occurs at the end of the abscissa. The judgment of desirability is 0 for least desirable and 1 for most desirable; it is plotted on the ordinate. For example, the top graph indicates, in words, that "more is preferred", while the next graph indicates that "less is preferred". The relative weights represent the relative importance given to each factor in making a judgment. The relative weights sum to 1 and the relative weight for each factor is scaled in terms of a number from 0 to 1.

For a given set of scenario outcomes the value of a particular variable and the corresponding function form determine the value of desirability for that variable. A linear model of judgment making is assumed. Quantitatively, each desirability value is multiplied by its associated weight and is added to the other desirability values (each multiplied by its weight in the same way); the sum gives the overall level of preference for that set of outcomes. Thus

$$\hat{Y} = \sum_{i=1}^{n} \beta_i [a(X_i - \bar{X}_i)^2 + bX_i + c] + e \qquad (i = 1, 2, \ldots, n) \tag{1}$$

where $\hat{Y}$ is the prediction of the individual's judgment, β_i are the weights associated with each factor, X_i are the numerical values of the n factors, $\bar{X}_i$ are the mean values of the n factors, a, b, and c are constants, and e is a random error term.

An additional positive feature of this theoretical framework, which is used to define the social values and to link them to a technical analysis, is the ability of the framework to quantify presently imprecise or unquantified, yet very important, aspects of the problem for inclusion in the quantitative assessment model. This was done for the side-effects variable. It was demonstrated in earlier work that the inclusion of such a variable was necessary. Thus the side-effects variable is a combined rating of how large the negative side effects of each policy option are thought to be; the larger the rating given, the greater the negative side effects expected. Side effects are effects other than the positive effect of reducing air pollution for which the policy option is proposed. The side-effects variable indicates that each time a policy option is turned on a cost is incurred which must be weighed against the benefit of reduced air pollution or energy use resulting from that option. The cost is part of the trade-off that must be made.

4 POLICY ANALYSIS OF THE SCENARIOS

For a more realistic evaluation of the scenarios and to demonstrate the rapid analysis made possible by our techniques, we determined the weights and function forms for the

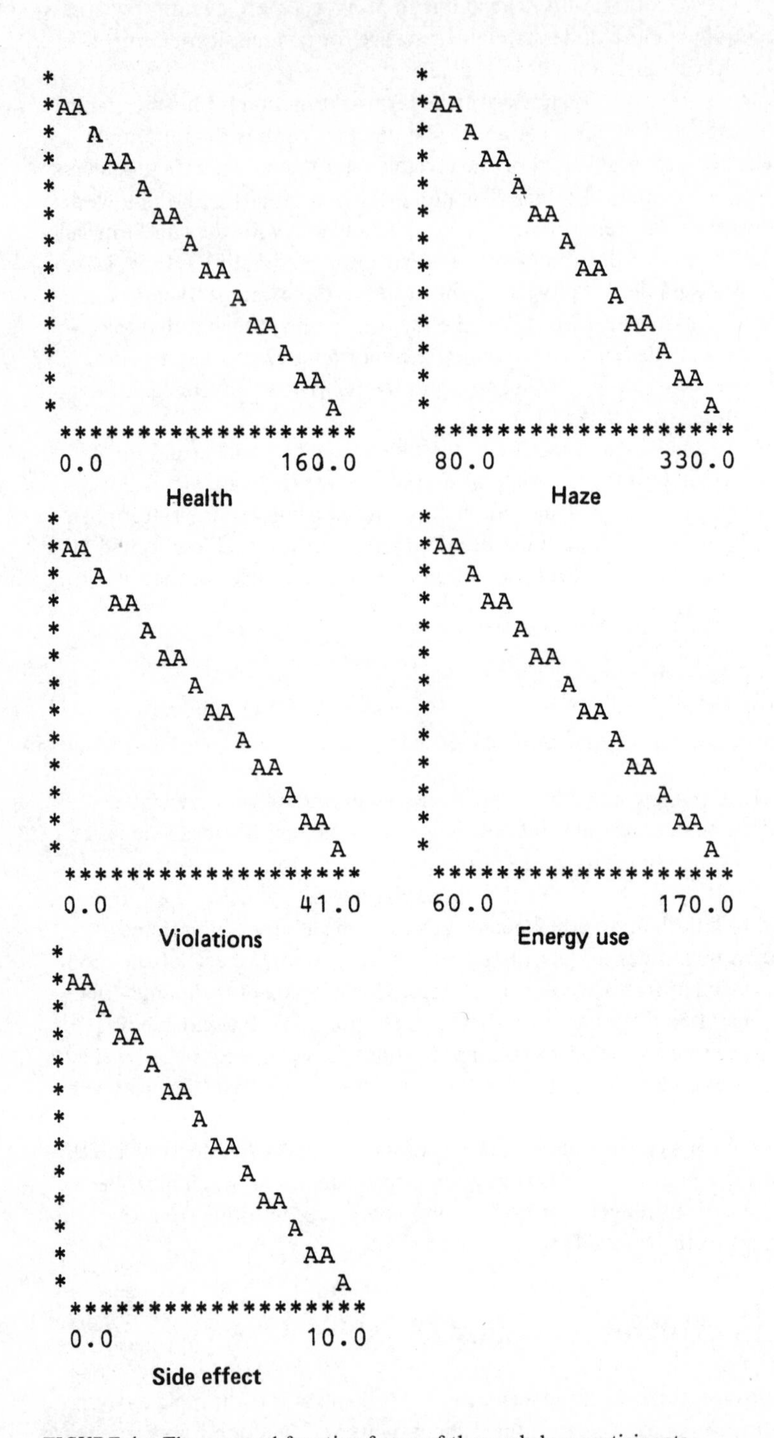

FIGURE 4 The averaged function forms of the workshop participants.

workshop participants one day and presented the analyzed results the next. We used a composite average set of weights and function forms as the set of social values for the analysis. This resultant set of function forms is shown in Figure 4 and the corresponding weights in Table 3. Note the extreme importance given to health effects.

TABLE 3 The averaged weights given by the workshop participants to the five effect categories.

Effect category	Weight	Function form[a]
Health	0.47	NEGLIN
Haze	0.11	NEGLIN
Violations	0.11	NEGLIN
Energy use	0.18	NEGLIN
Side effects	0.14	NEGLIN

[a]NEGLIN stands for negative linear.

For the analysis of the scenarios we developed software packages to interrogate and display the scenarios and their rankings in several different ways:

(1) individual rankings by year of the policy packages (switches listed as ON or OFF) and the scenario outcomes;
(2) rankings of the different policy packages as a function of time;
(3) comparison of the preference values for the outcomes from the scenarios for high, medium, and low growth for a given year and the same policy-option package.

Display (3) is one step towards addressing the effect of uncertainty in the decision-making arena.

4.1 Individual Annual Rankings

The top-ranked policy-option package for 1987 is given in Table 4. The social values used for the rankings are those of the workshop participants. The participants represent a select group of people who are actively involved in the air-quality problems of Denver as

TABLE 4 The first-ranked policy-option package for 1987 and its ranking.

Intervention pattern		*Outcomes*	
Car pooling	ON	Health effects	0% current levels
No-drive day	OFF	Haze	104% current levels
Mass transit	OFF	Standards violations	1 per season
Inspection–maintenance	OFF	Energy use	77% current levels
Diesel-car ban	OFF	Side effects	1 out of 10
Staggered work hours	OFF		
4-day work week	OFF	Rating	19.03 out of 20

TABLE 5 The second-ranked policy-option package for 1987 and its ratings.

Intervention pattern		*Outcomes*	
Car pooling	OFF	Health effects	0% current levels
No-drive day	OFF	Haze	109% current levels
Mass transit	OFF	Standards violations	1 per season
Inspection–maintenance	OFF	Energy use	84% current levels
Diesel-car ban	OFF	Side effects	0 out of 10
Staggered work hours	OFF		
4-day work week	OFF	Rating	18.96 out of 20

planners, interested citizens, and decision makers. The use of their social values for the workshop analysis is illustrative and is not intended to represent the actual social values of the Denver area. Car pooling is the only policy that is implemented. The second-ranked policy-option package for 1987 is given in Table 5. This is the do-nothing package since no policies are activated. Note how close the ratings of these two option packages are: 19.03 and 18.96 respectively. There is effectively no difference; they are practically equally desirable. A rating of 20 is perfect, but that is never achieved.

TABLE 6 The first-ranked policy-option package for 2000 and its rating.

Intervention pattern		*Outcomes*	
Car pooling	ON	Health effects	0% current levels
No-drive day	OFF	Haze	167% current levels
Mass transit	OFF	Standards violations	0 per season
Inspection–maintenance	OFF	Energy use	76% current levels
Diesel-car ban	OFF	Side effects	1 out of 10
Staggered work hours	OFF		
4-day work week	OFF	Rating	18.54 out of 20

The first-ranked option package for the year 2000 is given in Table 6. The option package with only car pooling implemented is still the most preferable, with a small decline in rating to 18.54. This result for the year 2000 can be understood by looking at the base-case numbers in Table 2 and at the weights in Table 3. Health effects, the most heavily weighted, are zero in 2000 for both the car-pooling and the do-nothing options. The small reductions in haze and energy use offset the negative aspects of the side effects anticipated from the implementation of car pooling. Of the seven options, the car-pooling option had one of the lowest levels of anticipated side effects. The implementation of an additional policy option did not produce sufficient reduction in the outcomes of lesser weight (health is already zero) to offset the undesirability of the anticipated side effects.

The first-ranked option package for the year 2020 is given in Table 7. Here three policy options have been implemented: car pooling, inspection–maintenance, and a ban on diesel cars. More drastic measures were required to reduce the health effects to zero. Note that the overall rating has now dropped significantly to 15.33 and the side-effects variable is at the level of 5 out of 10. The workshop participants were "becoming" less

TABLE 7 The first-ranked policy-option package for 2020 and its rating.

Intervention pattern		*Outcomes*	
Car pooling	ON	Health effects	0% current levels
No-drive day	OFF	Haze	178% current levels
Mass transit	OFF	Standards violations	0 per season
Inspection–maintenance	ON	Energy use	142% current levels
Diesel-car ban	ON	Side effects	5 out of 10
Staggered work hours	OFF		
4-day work week	OFF	Rating	15.33 out of 20

and less satisfied with the overall situation, even though this was the option package that they most preferred of the set of option packages allowed. The improvement in visibility predicted as an outcome of a ban on diesel cars was significant and large enough to offset the greater side effects anticipated for this option. The two legislative actions (the no-drive day and the ban on diesel cars) were anticipated to have the worst side effects of all the options. Thus one begins to sense the many trade-offs involved and the ability of this formalism to make these trade-offs amenable to inspection and understanding.

4.2 Rankings Over Time

Another means of presenting the information which further elucidates what has been learned from consideration of individual years is to display the rankings over time. This has been done in Table 8. All the options that are ranked high between 1983 and 2000 rank very low in 2020 because of the emphasis on health effects. The top-ranked option in 2020 ranks very poorly up to the year 2000. The only option that ranks moderately well for all periods after 1980 is the inspection–maintenance option. Interestingly, the do-nothing scenario fares quite well until the year 2020. A different view of the policy-option packages is obtained in this display of rankings over time. Some policy packages that do not look very desirable when individual years are analyzed may look better when

TABLE 8 The ranking[a] over time of six policy-option packages.

	Rank									
Scenario[b]	1975	1980	1983	1985	1987	1990	1995	2000	2010	2020
DN	8	2	2	2	2	2	2	2	4	73
CP	13	6	1	1	1	1	1	1	1	54
MT	18	5	8	7	7	5	5	5	12	68
IM	20	10	12	11	11	11	11	14	11	14
ND	28	18	3	8	9	9	9	9	7	62
CP–IM–DC	67	62	55	52	50	49	44	40	10	1

[a]The total number of possible scenarios is 96; the highest ranking = 1, the lowest ranking = 96.
[b]DN, do-nothing option; CP, car-pooling option; MT, mass-transit option; IM, inspection–maintenance option; ND, no-drive day option; CP–IM–DC, option with car pooling, inspection–maintenance, and a diesel-car ban.

the policy packages are analyzed across years if a premium is placed on having a decision that will remain appropriate for a long period.

The analysis presented here is highly dependent on the scenario model and on what is or is not included in it. These conclusions are merely demonstrative of the technique. However, once it is known which variables are most heavily weighted by the decision makers or the public, more effort can be usefully expended to improve those components of the model. For example, measures of other health effects need to be included, to ensure that the full complement of possible effects and the different histories of different effects are addressed. The analysis also points out that there may be pitfalls in consistently emphasizing one variable as the most important. Preference rankings may then be swayed by large relative changes in a variable whose magnitude is quite small, e.g. changes of 100% in health effects when the health effects are predicted to be 3% of current levels. The analysis presented here can provide a framework for addressing such interpretive problems.

4.3 Uncertainties

A problem hindering attempts to bring formal systematic analysis to social-policy formation is the difficulty of coping with the uncertainties that are usually prominent in the process of policy formation. In the approach described here, uncertainties are addressed through sensitivity analysis. A recently developed software package can be used to address both (a) uncertainties concerning the appropriate social values that should be used in the policy evaluation and (b) uncertainties concerning projections of the likely effects or projections of driving variables. An example of each type of sensitivity analysis was presented at the September 1979 workshop.

4.3.1 Uncertainties in Social Values

The process of constructing models of social values appears to facilitate communication and understanding and therefore leads to the development of consensus or compromise concerning the choice of an appropriate set of values for deciding between policy alternatives. However, there may still be uncertainty about the relative weights that should be assigned within the set of values. Sensitivity analysis enables one to determine whether alternative weightings of the components of the set of social values lead to different desirability assessments of different policy packages. Such a sensitivity analysis was carried out for the workshop.

Given that the weights assigned by the workshop participants put a heavy emphasis on health effects and given the knowledge that visibility degradation need not be correlated with health effects, an alternative set of weights was defined to contrast with the weights assigned by the workshop participants. Both sets of weights are shown in Table 9; the A weights were assigned by the workshop participants and the B values are the alternative set of weights. The greatest emphasis in the alternative set is put on haze and energy use, which expresses the opinion that these variables are what people may actually react to in considering health effects. Thus group B represents a very different set of weights from group A chosen by the workshop participants.

For 1987 the first-ranked scenario for group B was found to be the same as that for group A. This is understandable because not much could be affected to produce much

TABLE 9 The alternative set B of weights compared with the set A of averaged weights given by the workshop participants.

	Weight		Function form[a]	
Effect category	A	B	A	B
Health	0.47	0.07	NEGLIN	NEGLIN
Haze	0.11	0.31	NEGLIN	NEGLIN
Violations	0.11	0.16	NEGLIN	NEGLIN
Energy use	0.18	0.27	NEGLIN	NEGLIN
Side effects	0.14	0.20	NEGLIN	NEGLIN

[a]NEGLIN stands for negative linear.

TABLE 10 The first-ranked policy-option package and its rating for the alternative set of weights for the year 2020.

Intervention pattern		*Outcomes*	
Car pooling	ON	Health effects	0% current levels
No-drive day	OFF	Haze	178% current levels
Mass transit	OFF	Standards violations	0% per season
Inspection–maintenance	ON	Energy use	142% current levels
Diesel-car ban	ON	Side effects	5 out of 10
Staggered work hours	OFF		
4-day work week	OFF	Rating	11.89 out of 20

difference by 1987. It is surprising, however, that for 2020 the top-ranked scenario for group B, given in Table 10, is the same as the top-ranked policy-option package for group A. It should be noted that the overall level of desirability of 11.89 for the CP–IM–DC option package for group B is quite low compared with 15.33 for group A for the same policy-option package. Both sets of social values have the same policy-option package ranked at the top, but the second set shows much less satisfaction with the "world" as represented by these outcomes and would most probably provoke a search for new alternative options to be evaluated.

4.3.2 Technical Uncertainty

It is common practice in simulation modeling to replace one set of assumptions with another or to replace the expected value with values one or two standard deviations on either side and to observe resultant changes in the outcome variables. The present sensitivity-analysis program also adopts this general approach, with two important exceptions. First, the sensitivity analysis describes the variance in projected outcomes due to uncertainty not in terms of the natural units of the outcome dimensions but rather in terms of the desirability associated with these outcomes by a particular system of social values. Second, the analysis program deals with all sources of technical uncertainty simultaneously, rather than on a dimension-by-dimension basis, in order to facilitate assessment and comparison of the relative contribution of each source of uncertainty to the uncertainty in the overall desirability of policy alternatives.

For simplicity, only the results from a sensitivity analysis for the combined uncertainty of population and economic growth are presented. The rates of economic and population growth have a large uncertainty in their projected values and are difficult to affect, even if the region is interested in affecting them. The policies considered in the workshops are only trying to mitigate the effects of population and economic growth on air quality, without addressing the growth itself or the uncertainty in that growth. Without trying to affect the growth itself, one can ask what is the effect of the uncertainty in population and economic growth rates on the desirability of various policy option packages? For 1987 there is no effect, of course. Comparisons of several policy packages for the year 2020 are shown in Table 11.

TABLE 11 The variation in desirability ratings due to uncertainty in the population and economic growth rates for 2020.

Scenario	Rating 0	1	2	3	4	5	6	7	8	9	10	11	12	13	14	15	16	17	18	19	20
DN											H			M		L					
CP													H		M		L				
ND													H		M		L				
MT												H		M			L				
IM														H		M	L				
CP–IM–DC															H		M	L			

H, high; M, mean; L, low; the other abbreviations are the same as in Table 8.

It can be seen immediately that not all policy-option packages have the same sensitivity in their desirability ratings given the same uncertainty in population and economic activity in 2020. Some policy-option packages are more robust than others with respect to this particular uncertainty. For the particular set of social values used by the workshop participants, the inspection–maintenance option appears to give the greatest hedge against a policy-option package becoming too undesirable owing to the uncertainty of future population and economic growth. These conclusions are strongly dependent on the relationships in the scenario model, on the particular outcome variables modeled, and on the social values used for the evaluation. The important contribution of this procedure is that it provides an entirely different way of viewing uncertainty – in the metric of social values.

5 CONCLUSIONS

Several conclusions can be drawn from the workshop experience. First, the overall conclusion is that the method of explicitly linking social values and technical assessments is very promising for environmental-policy analysis. The provision of good technical assessments remains as difficult as ever. Second, the analysis did reflect much of the dynamics of the value-system trade-offs in the evaluation of scenarios. Third, the workshop mode proved to be a very good vehicle for potential users to obtain an overview and a basic understanding of the entire process. The workshop also proved to be a good vehicle for researchers to receive feedback from potential users and interested parties in order to determine directions for the next phases of the research project.

The fourth conclusion is that the present scenario model is not complex or rich enough in its components and structure to be of much use in actually deciding between policy-option packages, although much insight into this process was gained. A greater number and variety of outcome variables and policy options are needed. This shows how important it is that the process should be adaptive and proceed through several iterations. The next phase will therefore focus on the inclusion of more options and more varied outcomes as well as the provision of more in-depth analysis of a few select areas.

The conclusions drawn from the scenario model for the workshop were very dependent on the indicators used in the scenario model as surrogates for a whole class of effects. This was particularly clear for health effects based on CO exposure only. The preferences calculated for different policy packages should change as the health-effects indicator becomes more complete, i.e. as it includes NO_x and oxidants, because these pollutants do not follow the same time trend of emissions as CO. For other indicators the surrogates may be perfectly adequate to represent a class of behavior. The methods described here show promise as a framework for testing the sensitivity of the preferences for policy options to suspected missing information.

USING MATHEMATICAL PROGRAMMING MODELS FOR COST-EFFECTIVE MANAGEMENT OF AIR QUALITY

R.J. Anderson, Jr.*
Mathtech Inc., Princeton, New Jersey (USA)

1 INTRODUCTION

To an economist, a particularly tantalizing prospect explicitly raised in two of the papers in this volume is that mathematical models could be used to improve the efficiency (i.e. to lower the cost) of air-quality management. Gustafson and Kortanek (1982) (this volume, p. 75) show how time-invariant mathematical models of the costs of emissions controls can be combined with time-invariant models of the diffusion of emissions to form a mathematical programming model. This model represents the management problem of finding a set of emissions limitations (or emissions standards) that achieves a given ambient air-quality level at least cost. Melli et al. (1982) (this volume, p. 173) show how this programming model can be extended to incorporate time-varying phenomena, and to calculate cost-effective programs of real-time pollution control.

The basic idea that the problem of finding cost-effective control programs for emissions can be formulated as a mathematical programming model has been applied in several empirical investigations in the United States. The findings of these studies show uniformly that emissions limitations set on the basis of programming model solutions would be far less costly and no less effective than are the limitations typically set by US air-quality management authorities.

The results of a recent study (Anderson et al. 1979) vividly illustrate the potential magnitude of the savings that could be realized. Table 1 shows the estimated costs of meeting a prospective short-term NO_2 standard of 250 $\mu g/m^3$, maximum 1 h average, in the US city of Chicago, under a number of alternative management strategies. Only strategies which pertain to the control of stationary sources were considered. Mobile source controls to be applied under the US Clean Air Act were taken as given.

The top row of Table 1 shows what the situation would be if no emission controls were applied to stationary sources. Control costs would be zero, but the ambient standard would be exceeded at 104 of the 270 receptor sites considered in the model. The Least Cost control strategy, according to our calculations based on a mathematical programming

*Present address: IIASA, A-2361 Laxenburg, Austria.

TABLE 1 Optimization analysis results for Chicago (1978 dollars) (source: Anderson et al. 1979).

Strategy adopted	Number of sources controlled	Reduction in NO_x emissions (10^3 lb/h)	Number of receptors in violation of standard	Annual control costs (10^6 \$/y)
No control baseline	0	0	104	0
RACT	797	37	80	23
Least Cost	96	5	0	21
Selective Emissions Control	742	29	0	94
Simple Rollback				
90%	797	106	0	254
80%	792	104	6	243
Maximum Feasible Control	797	106	0	254

model, would entail an annual outlay of ca. US \$21 million per year to meet the ambient standard. Of more than 700 individual point emission sources considered, only 96 would have to apply controls to meet the ambient standard.

The lower rows of Table 1 report the results obtained when a number of alternative air-quality management strategies commonly applied in the United States today were simulated. The RACT strategy applied what is termed (in the US Clean Air Act) "Reasonably Available Control Technology" to all point sources*. As can be seen from the results reported, RACT fails to meet an ambient standard of 250 $\mu g/m^3$ (note that 80 of the 270 receptor points remain in violation) and costs \$23 million/year to implement, or \$2 million/year more than the Least Cost strategy.

The Selective Emission Control strategy represents what would be considered in the US today to be a relatively sophisticated air-quality management strategy. This strategy, as can be seen in Table 1, controls a high proportion of the sources in Chicago and would cost about \$94 million/year. This is over four times as much as control would cost under the Least Cost strategy identified by solution of our mathematical programming model! It should be remembered that the SEC strategy roughly represents what US air-quality managers consider to be a very efficient management strategy.

The remaining strategies for which results are reported in Table 1 (i.e. Simple Rollback, and Maximum Feasible Control) reflect what are generally conceded to be the crudest management approaches applied in the US today. The Maximum Feasible Control strategy applies the most effective known control technology to each of the point sources in Chicago. Simple Rollback requires that each source roll back (i.e. reduce) its emission in the same proportion that ambient concentrations exceed ambient standards, with an adjustment for so-called "background" concentrations. For example, if observed ambient concentrations were 280 $\mu g/m^3$, and if the ambient standard were 250 $\mu g/m^3$, then

*The US Clean Air Act requires all emissions sources in areas failing to meet ambient standards within a reasonable period of time to apply "reasonably available technology". The law states that such technology is to be economically affordable. The precise definition of RACT is left up to the Environmental Protection Agency, which has not yet defined RACT for NO_x sources. Thus, the results reported for RACT in Table 1 represent our rough estimate of what EPA might define RACT to be.

each source would be required under Simple Rollback to roll back emissions by 10.9%, according to the formula

$$\Delta A/A = (\Delta E/E) - (a\Delta E/AE)$$

where A is ambient concentrations, a is background, and E is emissions.

It is clear from the results reported in Table 1 that these latter two air-quality management strategies – i.e. Simple Rollback and Maximum Feasible Control – are far more costly than that required to meet ambient standards. Our results suggest that these strategies – which are sometimes employed in the US to manage air quality – may be ten times more costly than the Least Cost strategy.

Results like those described above certainly appear to indicate that use of mathematical programming models may contribute to an enormous improvement in the efficiency and effectiveness of air-quality management. It seems likely that the costs of abatement under currently practiced management strategies could be reduced by a factor of three or more.

2 OBJECTIVES

This paper assumes that mathematical programming models can and will substantially lower the cost of meeting ambient air-quality goals. Its purpose, starting from this premise, is to explore exactly how mathematical programming models should be used to secure the greatest possible improvement in the cost-effectiveness of air-quality management.

This, at first, certainly sounds like a curious and perhaps not very fruitful undertaking. It seems apparent after all, that air-quality management authorities should simply use the Least Cost solutions of programming models to set emissions standards for each source falling under their authority. Matters, however, are not that simple. In particular, the following analysis will show that there are better ways of using mathematical programming models. These involve the indirect use of model results to set economic disincentives – instead of emission standards – to bring about the reductions in emissions needed to achieve air-quality goals.

In addition to direct use of model results to set emissions standards, two indirect modes of use of model results are examined. The first indirect approach entails the use of model results to calculate shadow prices on emissions which would (in economies in which prices are used as incentives to guide production and consumption decisions) induce sources to apply the requisite emissions controls. These prices, or "emissions charges", would create financial incentives for managers of emissions sources to take steps to reduce their emissions. (This is explained further in Section 4.) The second indirect approach or mode of use entails the creation of transferable emissions permits, each of which would entitle its holder to emit a stated quantity of a specified pollutant. In this application, the programming model would be used to allocate permits among emitters on the basis of their offers to buy and/or sell permits. In effect, this approach entails the creation of a market in which emissions can be traded between pollution sources. Prices would be established by these transactions.

How could these indirect modes of use be preferable to simple direct use of the model results to set emissions standards? The reader familiar with duality theory in mathematical programming, for example, may protest that the use of the model to set emissions charges, as outlined above, is (almost) equivalent to the use of the model to set emissions standards. What then, would be the advantage of an indirect method of use?

The answer is that, in a mythical world of perfect information and costless decision making, there would be no advantage. The three modes of use (i.e. direct use to set standards, indirect use to set charges, and indirect use to allocate permits) would be precisely equivalent. In the real world, however, in which information is far from perfect and decision making far from costless, there would be differences between the results of the three approaches. As noted above, it will be shown that the indirect modes of use would tend to result in lower costs (see Section 6 for details) of implementing air-quality goals.

While the analysis used to establish this result may seem abstract, the conclusion could hardly be more practical. Many papers in this volume validate the basic premise of this argument: the information on which air-quality management decisions must be based is imperfect.

3 PLAN OF THE PAPER

In Section 4 an explanation is given of the basic economic concepts which underly the use of emissions charges or transferable permits in lieu of emissions standards as a means to reduce emissions. These concepts are well-known to most economists. They are discussed here to insure that readers from other disciplines and/or readers from economies that do not use prices as guides to production and consumption decisions are provided with the background needed to understand the present analysis.

Section 5 provides a brief review of the formulation of mathematical programming models of cost-effective air-quality management. This section also indicates briefly how such models could be used either to manage air quality directly through prescription of emission standards or indirectly through calculation of emissions charges or permit allocations and prices.

Section 6 contains an analysis of alternative modes of use of mathematical programming models. As noted above, it will be shown that the indirect use of such models would result in lower expected costs of managing air quality than would direct use of models to set emissions standards. Some results will be also derived concerning the certainty with which ambient air-quality goals would be obtained under these alternative uses. The general conclusion is that on both counts the use of the model to administer a system of transferable emission permits is the best of the three modes of use.

Section 7 offers some additional comments concerning the evaluation of alternative approaches for managing air quality. These comments – although less than rigorous – suggest additional reasons for believing that a transferable emissions permit approach to air-quality management offers the best practical prospect for efficient and effective management of air quality.

4 THE BASIC ECONOMICS OF EMISSIONS CHARGES AND TRANSFERABLE EMISSIONS PERMITS

As stated in Section 2, two different indirect approaches to air-quality management will be considered. Under one approach, the air-quality authority sets prices, which we call "emissions charges", that are levied upon sources' emissions. In the other approach, a fixed number of transferable permits is issued which sources may buy and sell as they see fit. This section explains when and why these approaches could be expected to be effective measures for reducing pollution.

4.1 Assumptions

The basic premise on which the emissions charges and transferable emissions permits approaches rest is that making pollution emissions costly will bring about a reduction in emissions. The assumption is that decision makers (e.g. managers of establishments which emit pollutants) would respond to financial incentives by seeking production methods that minimize costs, including liabilities for emissions charges or for payments for emission permits.

Two comments on this assumption are warranted. First, in some economies, financial incentives by design play little or no role in decision making concerning the allocation of resources. It follows that emissions charges or transferable emission permits would not be applicable approaches in these economies. The analysis presented in this and subsequent sections is relevant in these situations only insofar as the reader may wish to hypothesize that such economies might consider using prices as a central part of the planning and resource allocation process. The analysis presented in Section 4.2 demonstrates that there are situations in which this change in economic policy could improve resource allocation.

Second, even in economies in which financial incentives do play a role in resource allocation decisions, decision makers demonstrably are motivated by additional considerations. These other considerations are often in direct conflict with cost minimization. Consider, for example, the company executive who chooses to fly first class at company expense, and to have a 500 m^2 office when a 250 m^2 office would be adequate, and so forth.

It is clear then that the assumption of cost-minimizing behavior is only an approximation of actual behavior. Nonetheless, it is one whose empirical implications correspond tolerably well with behavior observed in many sectors of economies which make use of price incentives, and it is one which will be made here for purposes of conducting the analysis*.

4.2 An Example

With this basic assumption in mind, a simple example will clarify how prices can be used to reduce emissions and, in the case of transferable (or marketable) emissions permits,

*The conclusions reached that charges or permits are effective in inducing reductions in emissions do not depend on strict cost-minimizing behavior. However, the conclusion that indirect approaches would result in minimization of costs would need to be qualified.

how these prices might be established. For expository purposes, let us consider a hypothetical region in which there are two sources of particulate emissions. Basic data describing the characteristics of these sources are shown in Table 2.

TABLE 2 Basic data for hypothetical emission sources.

Emission control technologies	Source 1		Source 2	
	Emissions	Costs	Emissions	Costs
No controls	1000	0	200	0
Cyclone	500	10	100	5
Cyclone–precipitator	130	50	30	75
Cyclone–precipitator–baghouse[a]	100	100	0	300

[a] A building containing filters for removing particulate matter from gaseous emissions.

It will also be assumed, as discussed above, that the managers of these sources are sensitive to price incentives. In particular, it will be assumed that they seek to manage their facilities in such a way as to minimize the cost of operating the facility.

In the absence of any regulation or financial inducement to the contrary, the best course of action with respect to pollution control for each of the managers of our hypothetical sources is clear: the manager of Source 1 will use no emission controls, thereby adding nothing to his costs, and 1000 units will be emitted. Similarly, Source 2 will emit 200 units, and incur no control costs.

Suppose now that the air-quality management agency established a charge of five monetary units per unit of emission. The choices facing each of the managers of the sources shown in Table 2 would then be as reported in Table 3. The situation facing these

TABLE 3 Hypothetical data on costs and charge liabilities.

	Source 1				Source 2			
	Emissions	Costs	Charges	Total	Emissions	Costs	Charges	Total
No control	1000	0	5000	5000	200	0	1000	1000
Cyclone	500	10	2500	2510	100	5	500	505
Cyclone–precipitator	130	50	650	700	30	75	150	225
Cyclone–precipitator–baghouse	100	100	500	600	0	300	300	300

managers is now obviously different. If both managers persisted in not controlling their emissions, their costs would be quite high. Source 1's costs plus liabilities for charge payments if its emissions were not controlled would be 5000 monetary units; Source 2's costs plus charge liabilities would be 1000 monetary units.

Table 3 shows that the managers of both sources could reduce their total costs plus charge liabilities by reducing their emissions. For Source 1, the lowest total of control costs and charge liabilities is obtained (in this example) by using a cyclone in conjunction with a precipitator, and a baghouse. For Source 2, the lowest total cost option is a cyclone plus precipitator. Source 1 is able to reduce its total control costs plus charge liabilities to 600 monetary units, while Source 2 can reduce its total to 225 monetary units.

In effect, each incremental application of more effective emission-control technology reduces the sources' liabilities for payments of emission charges by reducing their emissions. Additional emission reduction pays off as long as the amount of the reduction in charge liabilities obtained through incremental emission reduction is greater than the additional control costs incurred to obtain the reduction.

This example can also be used to illustrate how a transferable emissions permit approach (as opposed to an emissions charge approach) works to establish a price on pollution. Let us suppose that, initially, only Source 1 existed in our hypothetical region and that the air-quality management authority issued 130 permits, each of which entitled its holder to emit 1 emission unit. Let us also suppose (purely to simplify matters) that the agency has simply issued all these permits to Source 1.

Now suppose that Source 2 wishes to locate in the region. In order to do so it will consider making some arrangement with Source 1 to secure emission permits. Negotiations would take place. Source 1 realizes that it can sell up to 30 permits and still remain in business*. If it did so, its permit holdings would entitle it to emit only 100 units. It would therefore have to install additional controls (in our example it would have to add a baghouse), at an additional cost of 50 monetary units (i.e. 100 monetary units less its present costs of 50 monetary units). Clearly then, it would be willing to sell 30 permits to Source 2 only if Source 2 paid at least 50 monetary units, i.e. 1.67 units per permit.

The maximum amount that Source 2 would pay for 30 permits, as can be seen from Table 2, is 225 monetary units. This is because, if Source 1 refuses to sell Source 2 any permits, Source 2's costs would be 300; if Source 2 obtained 30 permits, its control costs would be 75. Source 2 thus would save control costs of 225 monetary units by acquiring 30 permits. This, then, is the maximum that Source 2 would be willing to pay for 30 units. The price at which the exchange would be made would lie between the extremes of Source 1's minimum acceptable price per permit (1.67 monetary units) and Source 2's maximum acceptable price (7.50 monetary units per permit).

These simple examples show how prices can be used in place of direct emissions controls as a tool of air-quality management policy. If managers of pollution sources seek to minimize costs, then indirect management of pollution via prices will be an effective method of reducing emissions.

5 FORMULATION

Let us now consider in more detail the three alternative modes of use of mathematical programming models of cost-effective air-quality management (i.e. to set emissions standards, to set emissions charges, and to set permit prices and quantities in response to emitters bids) mentioned in Section 1.

*We assume that emissions cannot be made lower than 100 units if the source is to remain in business. To keep our example as simple as possible, we do not consider the possibility that a source would sell all of its permits and cease operations in a region. In practice, of course, this could and would happen in some instances.

5.1 Notation and Assumptions

For this purpose, it is helpful to introduce some notation, and simplifying assumptions. The basic assumptions are as follows:

(a) We consider only the case of a single receptor (i.e. location at which air quality is to be managed).
(b) $C_i(e_i, u_i)$ represents the ith source's cost function, where e_i represents its quantity of emissions, and u_i is a random variable representing the control agency's uncertainty about the source's costs.
(c) $d_i(v_i)$ represents the uncertain coefficient of proportion relating units of emissions from the ith source to ambient concentrations at a receptor. "v_i" is a random variable representing the agency's uncertainty about the diffusion relationship.
(d) s represents an ambient standard.
(e) $E(v_i, u_j) = E(v_i)E(u_j) = 0$, where E is the mathematical expectation operator; i.e., the random variables v_i and u_j are statistically independent for all i and j.
(f) In view of uncertainty in the diffusion relationship, the agency seeks control policies giving ambient concentrations equal to the ambient standard.
(g) The sources know their control costs with certainty.

These assumptions are stronger than strictly required to be able to derive the results obtained below. For example, it is not necessary to assume [as in (g) above] that sources are certain about their costs; it is sufficient to assume that they are less uncertain about them than the agency.

5.2 Specification of Alternative Modes of Use

The objective is to compare three different uses of mathematical programming models of cost-effective air-quality management.

Under the first use, i.e. direct use of the model to establish emission standards for sources, the agency sets standards so as to minimize expected costs, subject to achieving expected ambient concentrations equal to the ambient standard. This problem can be formulated mathematically as

$$E\left\{\sum_{i=1}^{N_s} C_i(\hat{e}_i, u_i)\right\} = \min_{e_i} E\left\{\sum_{i=1}^{N_s} C_i(e_i, u_i)\right\} \tag{1}$$

subject to

$$E\left\{\sum_{i=1}^{N_s} d_i(v_i)e_i\right\} = s$$

The solutions $\hat{e}_i$ are the emissions regulations which minimize expected costs, subject to the ambient constraint, and

$$E\left\{\sum_{i=1}^{N_s} C_i(\hat{e}_i, u_i)\right\}$$

is the expected cost under this set of regulations.

Under the second use of mathematical programming models examined, the management authority uses the model to set emissions charges so as to minimize expected cost while attaining expected ambient concentrations equal to the ambient standard. $h_i(p_i, u_i)$ is the function relating the ith source's emissions to the charge level set. The random variable u_i is included, reflecting the fact that the management authority, since it does not know the source's cost function, cannot be certain about the source's response to any given emission charge. Then, the management authority's problem is to find the set of p_i such that

$$E\left\{\sum_{i=1}^{N_s} C_i[h_i(\hat{p}_i, u_i), u_i]\right\} = \min_{p_i} E\left\{\sum_{i=1}^{N_s} C_i[h_i(p_i, u_i), u_i]\right\} \quad (2)$$

subject to

$$E\left\{\sum_{i=1}^{N_s} d_i(v_i) h_i(p_i, u_i)\right\} = s$$

The third use of mathematical programming models under consideration requires the sources to bid to purchase permits from the agency. The agency is to sell no more permits to sources than would result in expected ambient concentrations just equal to the ambient standard. It will be assumed that each source bids the maximum amount it is willing to pay for permits. Each submits its bids in the form of a demand schedule for permits.

To see the relationship between control costs, willingness to pay for permits, and bids, note that the marginal cost savings to, e.g., Source i from purchase of an incremental permit is simply $C_i(e_i, u_i)$. This is the amount that it would be willing to pay for the last permit unit purchased, and, taken as a function of e_i, represents the bid function of Source i for permits. Assuming that the management authority does not wish to extract the highest possible price for permits, but rather to maximize the net value of (i.e. willingness to pay for) permits issued subject to the constraint that ambient standards be met on an expected value basis, it can be shown that the agency's problem is, given each source's bid function, to choose the number of permits to issue to each source, $\bar{e}_i$, (and the corresponding price to charge) such that

$$E\left\{\sum_{i=1}^{N_s} C_i(0, u_i) - C(\bar{e}_i, u_i)\right\} = \max E\left\{\sum_{i=1}^{N_s} C_i(0, u_i) - C(e_i, u_i)\right\} \quad (3)$$

subject to

$$E\left\{\Sigma\ d_i(v_i)e_i\right\} = s$$

which is precisely equivalent apart from the constant terms $E\{C(0, u_i)\}$ to

$$E\left\{\sum_{i=1}^{N_s} C_i(\bar{e}_i, u_i)\right\} = \min_{e_i} E\left\{\sum_{i=1}^{N_s} C_i(e_i, u_i)\right\} \qquad (4)$$

subject to

$$E\left\{\Sigma\, d_i(v_i)e_i\right\} = s$$

where $E\{\Sigma\, C(\bar{e}_i, u_i)\}$ represents expected total costs at the optimal permit allocation.

The most important thing to note about eqn. (4) is its similarity to eqn. (1). There is however one important difference. Equation (1) represents the problem faced by the agency before it has received bid schedules from the sources. But, under a permit approach, the problem to be solved by the agency begins *after* it has received bid schedules. This means that, under the assumption that firms are perfectly certain about their costs and base their bids upon them, the agency will be perfectly certain about control costs when it makes its permit allocation. This means that it can set an allocation that exactly minimizes costs.

6 ANALYSIS

The problem in this analysis is to evaluate the results of each of the three modes of model use described above. These results will be evaluated in terms of two criteria. The first criterion is the expected cost to which each approach leads. Other things being equal, a mode of model use which leads to lower expected cost would be preferable to one which leads to higher expected cost. The second criterion is the variance of expected ambient concentrations. All uses considered have been constrained to result in expected ambient concentrations equal to the ambient air-quality goal. This being the case, other things being equal, modes of model use that result in a relatively tight distribution of ambient concentrations around the ambient standard would be preferred. Accordingly, in the following analysis, each step is taken to produce manageable expressions for expected costs and for the variances of ambient concentrations. These expressions are then compared to rank the three modes of use considered in terms of the criteria adopted.

6.1 Some Useful Approximations

The starting point for our analysis is a quadratic approximation of the sources' cost functions. This is chosen because it is far easier to work with quadratic forms than with general functional forms, and because, in many instances, quadratic forms provide good approximations. The cost functions about the points $\hat{e}_i$ will be approximated by the functions

$$C_i(e_i,u_i) \cong a_i(u_i) + [C_i' + \alpha_i(u_i)(e_i - \hat{e}_i)] + C_i''(e_i - \hat{e}_i)^2/2 \qquad (5)$$

where the $a_i(u_i)$, and $\alpha_i(u_i)$ are random variables, and where the C_i' and C_i'' are constants.

It will be assumed that the $\alpha_i(u_i)$ have been standardized so that $E\{\alpha_i(u_i)\} = 0$ for all i. Note also that since u_i and v_j are independent for all i and j by assumption,

$$E\{\alpha_i(u_i)d_j(v_j)\} = E\{\alpha_i(u_i)\} \cdot E\{d_j(v_j)\} = 0$$

The basic approximations given in eqn. (5) and assumptions about the random errors imply several other approximations that will be useful in the analysis developed below. The most important of these are as follows:

$$\mathrm{d}C_i/\mathrm{d}e_i = C_i'(e_i,u_i) \simeq [C_i' + \alpha_i(u_i)] + C_i''(e_i - \hat{e}_i) \qquad (6)$$

$$E\{C_i'(\hat{e}_i,u_i)\} \simeq C_i' \qquad (7)$$

$$E\{C_i''(e_i,u_i)\} \simeq C_i'' \qquad (8)$$

Note that eqns. (7) and (8) provide an interpretation of each of the fixed coefficients which appear in the basic approximating equation.

6.2 Step-by-Step Analysis

We now have all of the basic ingredients and relationships required to explore the relative expected costs and certainty of attainment of ambient standards under the three modes of model use. There is a fair amount of manipulation and substitution in deriving the results to be obtained, so it may be useful here to comment on the steps involved in the analysis.

We will begin by comparing the use of the model to set emissions standards [as described in eqn. (1)] to the use of the model to set emissions charges [as described in eqn. (2)]. The first step in this comparison is to derive an explicit approximate expression for the functions $h_i(p_i,u_i)$, which give the distributions of emissions levels (as perceived by the emissions charge-setting agency) resulting from any given set of charges, p_i. Which levels of emissions actually arise depends, of course, on the value taken by the u_i, which is known under our assumptions only by the sources.

The second step of the analysis is to substitute the expressions derived for $h_i(p_i,u_i)$ into the equations representing the problem of finding a set of prices to minimize expected cost [eqn. (2)], and to find an expression for the values of the p_i, denoted $\tilde{p}_i$, which solve this problem.

The third step is to take the resulting expression for $\tilde{p}_i$, substitute it into the approximate expression for cost [i.e. eqn. (5)], and evaluate expected costs. This yields an estimate of expected costs which will be compared with a similar expression for expected costs evaluated at $\hat{e}_i$. This comparison will show that the expected total cost of air-quality management achieved by using the model to set emissions charges is lower than that achieved by using the model to set emissions standards.

The final step in comparing these two alternative modes of model use is to compare the variances of the resulting distributions of ambient air quality. Analysis will show that use of the model to set emissions standards results in a smaller variance in ambient concentrations than does use of the model to set emissions charges, with the size of the difference in the variances related to the same factors which tend to give use of the model to set charges an expected cost advantage. That is, the larger the expected cost advantage that a charges approach has over an emissions standards approach, the greater the dispersion of the distribution of ambient concentrations about the ambient standard level.

Our comparison of the results of using the model to facilitate operations of a market in transferable emissions permits to the results of using the model to set emission standards or emissions charges follows a sequence of steps similar to that outlined above. The analysis will show that the expected cost of a transferable permit mode of use, under our assumptions, is precisely equal to the expected cost of the emissions charges mode of use, and hence, is less than the expected cost of the emissions standard mode of use. It will be shown that the dispersion of the distribution of ambient concentrations resulting from a transferable permit approach is the same as that which would result from an emissions standard approach, and is less than that resulting from using the model to set emissions charges.

Proceeding according to the strategy outlined above, we begin by deriving an approximate expression for the functions $h_i(\cdot)$ which relate sources' resulting emissions levels to charges levels. We know that for the ith source to minimize costs when faced with an emissions charge p_i and state of nature u_i, emissions will be adjusted to the point where

$$-C_i'(e_i,u_i) = -C_i'[h_i(p_i,u_i)\ ,\ u_i] = p_i \tag{9}$$

That is, it will adjust emissions to the point where its incremental cost saving from increasing emissions is just equal to its incremental charge liability from increasing emissions*.

Substituting our approximate expression for the derivative of the cost function into eqn. (9) we obtain

$$-[C_i' + \alpha_i(u_i)] - C_i''[h_i(p_i,u_i) - e_i] \simeq p_i$$

which, after rearrangement, yields the following approximate expression for the functions $h_i(\cdot)$

$$\tilde{e}_i(u_i) = h_i(p_i,u_i) \simeq \hat{e}_i - \frac{p_i + C_i' + \alpha_i(u_i)}{C_i''} \qquad (i = 1, 2, \ldots, N_s) \tag{10}$$

These functions give the cost-minimizing levels of emissions that will be chosen by each source at any charge level. The random term $\alpha_i(u_i)$ which appears in each equation reflects the fact that the agency does not know with certainty what level of emissions each source will pick, because it does not know each source's costs with certainty.

The next step is to find the set of charges that minimizes expected costs by substituting eqn. (10) into the approximating expression for costs, and to evaluate approximate expected total costs. When this has been carried out we obtain

---

*For additional explanation, see the numerical illustration presented in Section 4.2.

$$E\left\{\sum_{i=1}^{N_s} C_i[h_i(p_i, u_i), u_i]\right\}$$

$$\simeq E\left\{\sum_{i=1}^{N_s}\left[a_i(u_i) + (C_i' + \alpha_i(u_i)) - \left(\frac{p_i + C_i' + \alpha_i(u_i)}{C_i''}\right) + \frac{C_i''}{2}\left(\frac{p_i + C_i' + \alpha_i(u_i)}{C_i''}\right)^2\right]\right\} \tag{11}$$

$$= \sum_{i=1}^{N_s}\left[E\{C_i(\hat{e}_i, u_i)\} + \frac{p_i^2 - C_i'^2}{2C_i''} - \frac{\sigma_i^2}{2C_i''}\right]$$

where

$$\sigma_i^2 = E\{\alpha_i(u_i)\,\alpha_i(u_i)\}$$

Using the approximate expression for expected total costs given in eqn. (11), we proceed to find an expression for the set of prices which minimize this expression and attain expected ambient concentrations approximately equal to the standard. This is done by minimizing eqn. (11) with respect to the p_i, subject to the expected air-quality constraint. The first-order conditions necessary for this are

$$\tilde{p}_i = \lambda \check{d}_i \qquad (i = 1, 2, \ldots, N_s)$$

$$\sum_{i=1}^{N_s} \check{d}_i E\left\{\hat{e}_i - \frac{\tilde{p}_i + C_i' + \alpha_i(u_i)}{C_i''}\right\} = s \tag{12}$$

where $\check{d}_i = Ed_i(v_i)$. Note that since expected ambient concentrations under the emission standard approach are required to be equal to the standard level, the constraint equation presented in eqn. (12) implies that $p_i = -C_i'$. This in turn implies [by substituting back in eqn. (10)] that at the optimal charge rates, the distributions of emission rates anticipated by the agency is given by the distributions of the random variables

$$\tilde{e}_i = \hat{e}_i - [\alpha_i(u_i)/C_i''] \qquad (i = 1, 2, \ldots, N_s) \tag{13}$$

and that expected costs under the optimal charge policy, as given by eqn. (11) evaluated at $p_i = -C_i'$, are

$$\sum_{i=1}^{N_s} [E\{C_i(\hat{e}_i, u_i)\} - (\sigma_i^2/2C_i'')] \tag{14}$$

6.3 Results

From eqn. (14) it is apparent that the expected cost obtained when the model is used to set emissions charges is less than that obtained when the model is used to set emissions standards. This can be seen by noting that the expected cost under the charges approach is equal to that under the emissions standards approach minus a positive term which depends upon the variances of the intercepts of the marginal cost function [i.e. the variances of the $\alpha_i(u_i)$]. The difference, Δ_C, between expected cost when the model is used to set emissions charges and that when it is used to set emissions standards is

$$\Delta_C \simeq \frac{1}{2} \sum_{i=1}^{N_s} (\sigma_i^2 / C_i'') \tag{15}$$

The expected cost advantage of the emissions charges approach comes at a price, however. The variance of ambient concentrations when the model is used to set emissions standards is

$$v_e = \sum_{i=1}^{N_s} \check{\check{d}}_i \hat{e}_i^{\,2} \tag{16}$$

where $\check{\check{d}}_i$ is the variance of $d_i(v_i)$. In contrast, the variance of ambient concentrations when the model is used to set emissions charges can be shown [by substituting the expression for the e_i in eqn. (10) into the constraint equation and evaluating the variance of the resulting expression] to be

$$V_i = \sum_{i=1}^{N_s} \check{\check{d}}_i \hat{e}_i^{\,2} + \check{\check{d}}_i (\sigma_i / C_i'')^2 \tag{17}$$

This is clearly larger than the variance of ambient concentrations under the emissions standards mode of use, and the difference between the two increases with increasing uncertainty about sources' costs.

Our comparison thus leads to the conclusion that the emissions charges approach would result in lower expected cost and less certainty about resulting ambient concentration than would the emission standards approach. The more uncertain are sources' cost functions, the larger the cost advantage of an emissions charges approach relative to an emissions standards approach, and the greater the ambient air-quality certainty advantage of an emissions standards approach relative to an emissions charges approach. Any choice between these two modes of use thus depends upon a weighing of control cost savings against the benefits of greater certainty concerning ambient air quality. Since we have no information on the benefits of ambient air-quality improvement, there is no basis for making this comparison here. It is simply pointed out that this trade-off has to be made in order to decide which of the modes of model use – to set emissions standards or to set emissions charges – is better.

It is a relatively simple matter to extend the analysis presented above to consideration of the expected cost and the dispersion of the distribution of ambient concentrations which would result from use of the model to operate a transferable emissions permit system. Under this mode of model use, given our assumptions that sources always know their exact cost functions and reveal them in bidding for permits, the agency allocates a number of permits to each source, $\bar{e}_i$, such that*

$$C_i'(\bar{e}_i, u_i) = \lambda d_i \tag{18}$$

Setting our approximate expression for incremental costs in eqn. (6) equal to the right-hand side of eqn. (18) and solving for e_i, we obtain

$$\bar{e}_i \simeq \hat{e}_i - \frac{\lambda d_i + C_i' + \alpha_i(u_i)}{C_i''} \tag{19}$$

Note the striking similarity between eqns. (19) and (10). Indeed, they are precisely the same equation, and a full analysis of the first-order conditions leading to eqns. (18) and (19) paralleling our analysis above leads to the conclusion that the optimal allocation $\bar{e}_i$ of permits under the transferable permit approach is

$$\bar{e}_i = \hat{e}_i - [\alpha_i(u_i)/C_i''] \tag{20}$$

It follows immediately, by substitution back into the approximate expression for costs and taking expectations, that the transferable emissions permit approach enjoys the same expected cost advantage that is enjoyed by the emissions charge approach. That is, expected cost under the transferable emissions permit mode of model use is less than that under the emissions standards mode, by the amount Δ_C given in eqn. (15).

In computing the variance of the distribution of ambient concentrations resulting from the transferable emissions permit mode of use we proceed as for the emissions charge mode of use, with one very important exception. Under the marketable permit approach, the management authority learns the $\alpha_i(u_i)$ from sources through the bids submitted *prior* to making its decision. The $\alpha_i(u_i)$ are thus no longer random variables when the decision is made, and the variance of ambient concentration is given approximately by eqn. (16) – which gives the variance of ambient concentrations under the emissions standards approach.

This is a most interesting result. If the assumptions upon which it rests are accepted, it can be concluded that both the transferable emissions permit and emissions charges modes of use have a cost advantage over the emissions standards approach. It is also concluded that the variance of ambient concentrations under the transferable emissions permit approach is equal to that under the emission standards approach. Based on our two criteria, expected cost and the precision with which a policy attains ambient standard levels, the analysis thus leads to the conclusion that the transferable emissions permit approach is the best mode of model use of the three examined here.

---

*Equation (18) follows directly from the first-order conditions necessary for the problem stated in eqn. (4) above.

7 CONCLUDING REMARKS

It is necessary to add that the results described here are based upon a particular theoretical model and some definite assumptions. While the model and assumptions seem reasonable enough, it is certainly possible to reach contradictory conclusions using other assumptions. For example, if the quadratic approximation adopted in eqn. (5) were invalid, the results based upon it may also be invalid. Since there is really no way of knowing for sure whether or not the approximation is adequate, the conclusions stated above must be qualified.

Anderson et al. (1979) presented more-detailed investigations of alternative policies, and dealt with a number of additional considerations which bear on the choice among management policies, on a more pragmatic, less theoretical, basis. Interestingly, these considerations – all basically related to the fact that different parties with interest in air-pollution control each possess different information – also seem to point to the conclusion that a transferable emissions permit approach is best.

Among the specific considerations which lead to this conclusion are the following points:

(a) To implement an emissions charge system or emissions standard system, the air-quality management authority must take the initiative to acquire information about sources' control costs. This is a difficult and expensive undertaking if costs are to be determined with any degree of accuracy. Under a transferable permit approach, the agency does not need to take the initiative to acquire detailed cost data. These data are revealed by sources indirectly as they bid for the purchase of permits.

(b) A permit system self-adjusts to inflation and growth. Under a charge system or emissions standard system, the air-quality management authority has to adjust charge rates or standards to reflect this factor, making changes which would depend upon data which are uncertain and perhaps difficult to obtain.

(c) It would be relatively easy to provide for contingent future transfers of emissions permits. This would provide decision makers with direct means to hedge against uncertainties about the future. There is no similarly "natural" way to hedge against uncertainty under an emissions standards or an emissions charges approach.

REFERENCES

Anderson, R.J., Jr., Reid, R.O., and Seskin, E.P. (1979). An analysis of alternative policies for attaining and maintaining a short-term NO_x standard. Mathtech. Inc. Prepared for the Environmental Protection Agency, the US President's Council on Environmental Quality and the US President's Council of Economic Advisors.

Gustafson, S.-Å. and Kortanek, K.O. (1982). A comprehensive approach to air quality planning: abatement, monitoring networks, and real-time interpolation. In G. Fronza and P. Melli (Editors), Mathematical Models for Planning and Controlling Air Quality. Pergamon Press, Oxford. (This volume, p. 75.)

Melli, P., Bolzern, P., Fronza, G., and Spirito, A. (1982). The cost of a real-time control scheme for sulfur dioxide emissions. In G. Fronza and P. Melli (Editors), Mathematical Models for Planning and Controlling Air Quality. Pergamon Press, Oxford. (This volume, p. 173.)

A COMPREHENSIVE APPROACH TO AIR-QUALITY PLANNING: ABATEMENT, MONITORING NETWORKS, AND REAL-TIME INTERPOLATION

S.-Å. Gustafson
Royal Institute of Technology, Stockholm (Sweden) and Institut für Angewandte Mathematik, Bonn (FRG)

K.O. Kortanek*
Carnegie–Mellon University, Pittsburgh, Pennsylvania (USA)

1 INTRODUCTION

Let S denote a two-dimensional geographic region such as a city or country, where an arbitrary point x in S shall be denoted by two coordinates (x_1, x_2) with x_1 being the horizontal distance from a preselected origin and x_2 being the vertical distance from the origin. Assume that samples of the air are taken at a fixed point x in S, termed a receptor point, and that the concentrations of certain pollutants are measured. A list of conceivable pollutants could include sulfur oxides (SO_x), suspended particulate matter, carbon monoxide (CO), oxidants (O), and nitrogen oxides.

The results of the measurements, termed the air quality at point x, must be interpreted as time averages about a fixed time t:

$$\chi(x,t) = (1/T) \int_{-T/2}^{T/2} \tilde{\chi}(x,t+\tau)\mathrm{d}\tau$$

The true concentration function $\tilde{\chi}$ cannot be measured directly; only the average χ is available. The time interval T is fixed but can be of different magnitudes in different circumstances. For example, T may vary from 15 min to 1 month.

Assume that measurements are taken continuously over the year at a receptor point x and averaged over a time interval T. Many measured concentrations are generated; however, they vary irregularly from one time interval to the next, and thus the air quality cannot be defined by a single number. However, Larsen (1969) has established statistical laws

*Also College of Engineering Visiting Professor, Virginia Polytechnic Institute and State University, Blacksburg, Virginia (USA).

related to the log-normal distribution which make it possible to estimate the frequency of various concentration levels at a receptor point.

In spite of the uncertainty involved in air quality much work has been done on the design of permanent control strategies to meet ambient air-quality standards at their mean values. It is often difficult to estimate the tails of a probabilistic frequency curve with good accuracy. However, once a long-term abatement policy has been implemented it can be tested in order to ascertain whether the shorter-term frequencies of high concentrations actually meet short-term air-quality standards. Otherwise permanent controls must be supplemented with temporary measures. Alternatively, one could lower the permitted long-term mean concentrations further than prescribed by the standard in order to decrease the frequencies of high concentrations.

In this paper we briefly review an optimization approach to the determination of selective abatement policies which employs an atmospheric-dispersion model as a key input. Over the last ten years there has been a substantial improvement in the quality of mathematical source-oriented air-quality dispersion models (see, for example, Hrenko and Turner, 1975; and Turner and Hrenko, 1978).

An additional stage in air-quality management is the determination of a sampling network of fixed monitoring stations from which empirical air-quality data are to be generated. The empirically generated data provide evidence on whether an abatement policy is being implemented or not; they also provide an overall measure of the air quality throughout the region. We briefly report on a method of choosing "typical" sampling points, i.e. places where the concentrations reach their peaks or where measurements are needed in order to estimate concentrations in other parts of the region S where there are no measuring devices.

The latter question is one of how to interpolate concentrations off the network possibly even on a real-time basis. This brings us to the third stage of our approach to air-quality management.

In the closing sections we review the aggregated–averaged approximation method which objectively interpolates the air-quality distribution over all of S in terms of sparse measurement data generated by an existing air-quality monitoring network. The distinguishing feature of the method is that it combines empirical information from the monitoring network with atmospheric-dispersion functions to provide improved estimates of the concentration distribution over the entire region. Because of the large number of different "weather" states that affect the atmospheric dispersion of pollution, considerable computation is required, although most of this can be done in advance. Consequently the final interpolations computed from actual measured values only require very simple calculations, and these can be done on a real-time short-term basis.

2 MATHEMATICAL OPTIMIZATION MODELS

Several models for the estimation of the concentrations of pollutants have been proposed. Here we confine ourselves to the case of a single chemically-inert pollutant like SO_2. However, the arguments may be generalized to more complex situations. The models which are applicable for our problem have the following properties.

(1) The input consists of a source inventory (giving the strengths and positions of the sources) and climatological data (the frequencies of combinations of the wind directions, the wind speeds, the stability class, and the mixing heights).

(2) There are transfer functions which relate the emission rates of each source to its concentration contributions at all points of S.

(3) The principle of superposition is valid: if we divide the sources into two groups whose concentration contributions are defined by u_1 and u_2 respectively and let the emissions of the two groups be reduced by the fractions E_1 and E_2, then the remaining concentrations will be

$$(1 - E_1)u_1 + (1 - E_2)u_2$$

We shall assume that the superposition principle holds for all averaging times T. Hence it is true for the annual mean.

Now let Ψ be a given standard, where Ψ is a continuous function on S. A reduction policy is called feasible if the concentration does not exceed Ψ anywhere in S. The statement that an adopted strategy is feasible must be verified a posteriori using all available information. Suppose that there are n sources of a chemically-inert pollutant in a region and that we wish to control the emissions so that the sampling-period mean ground-level pollution concentration at each point in the region satisfies some standard. Furthermore, suppose that we wish to find a control policy whose total cost is minimal. Let us briefly review what we shall take to be the costs of abatement.

2.1 Cost Specification

There are several ways of stating the costs of reducing SO_2 emissions. If the SO_2 reductions involve the method of fuel desulfurization, then the costs can be stated as a function of the weight of the separated sulfur. However, in the comparison of fuel-switching abatement measures it is convenient to state the costs as functions of the sulfur content of the fuel. In the latter instance the costs could be simply the purchase price of various fuels.

Actually, under assumptions of normal cost-function monotonicity both methods are equivalent, for it is possible to restate costs per sulfur content of fuel as costs per ton of separated sulfur if the average heat demand for a stationary source is constant throughout the sampling period. We shall make a further simplification by stating costs as functions of the fraction by which a source reduces its output of pollutant.

2.2 An Optimization Governed by Least Cost and its Economic Duality Interpretations

Using diffusion modeling and a source inventory we can compute the functions $u_r(x)$, the mean pollutant concentration at point x due to source r before control, and $u_0(x)$, the mean pollutant concentration at point x due to all sources not under the control of the regulator. We also need the following definitions and data: S is the control region, e_r is the fraction by which source r is to reduce its output of pollutant (control variable), $G_r(e_r)$ is the cost of reducing output by e_r at source r, $\Psi(x)$ is the maximum mean concentration of pollutant to be permitted at point x, and E_r is the maximum amount we will require source r to reduce ($0 < E_r < 1$).

Since we have assumed that the pollutant is chemically inert, we can use superposition to find the pollutant concentration at each point x in S after reduction by $e_1, \ldots, e_n$:

$$\chi_e(x) = \sum_{r=1}^{n} (1 - e_r) u_r(x) + u_0(x) \tag{1}$$

If we define the excess pollution at x before reduction,

$$\phi(x) \equiv \sum_{r=0}^{n} u_r(x) - \Psi(x)$$

a least-cost strategy will be a solution to the following program, which, with its mathematical programming duality interpretations, has been studied by Gribik (1978).

*Program I**

Find $V_{\mathrm{I}*} = \min \sum_{r=1}^{n} G_r(e_r)$ from among $e_1, \ldots, e_n$ which satisfy

$$\sum_{r=1}^{n} e_r u_r(x) \geqslant \phi(x) \quad \text{(for all } x \text{ in } S)$$

$$0 \leqslant e_r \leqslant E_r \quad (\text{for } r = 1,2,\ldots,n)$$

In the following we shall assume that Program I* satisfies four regularity assumptions:

(i) $G_r(\cdot)$ is a continuously differentiable increasing convex function for $r = 1,2,\ldots,n$.
(ii) $u_r(\cdot)$ is a continuously differentiable function for $r = 1,2,\ldots,n$, as is $\phi(\cdot)$.
(iii) S is a closed and bounded set.
(iv) There exist $\hat{e}_1, \ldots, \hat{e}_n$ such that

$$\sum_{r=1}^{n} \hat{e}_r u_r(x) > \phi(x) \quad \text{(for all } x \text{ in } S)$$

$$0 < \hat{e}_r < E_r \quad (\text{for } r = 1,2,\ldots,n)$$

Under assumptions (i)–(iv) Program I* will have optimal solutions; let $e_1^*, \ldots, e_n^*$ be one such solution. Also, we can show that the following program is dual to Program I*.

Program II

Find

$$V_{\mathrm{II}} = \max \sum_{x \in S} \lambda(x)\phi(x) - \sum_{r=1}^{n} \lambda_r E_r + \left[\sum_{r=1}^{n} G_r(e_r^*) - \sum_{r=1}^{n} e_r^* \, \mathrm{d}G_r(e_r^*)/\mathrm{d}e_r \right]$$

from among functions $\lambda(\cdot)$ on S and scalars $\lambda_1, \ldots, \lambda_n$ which satisfy

$$\sum_{x \in S} \lambda(x) u_r(x) - \lambda_r \leqslant \mathrm{d}G_r(e_r^*)/\mathrm{d}e_r \quad (\text{for } r = 1,2,\ldots,n)$$

$$\lambda(x) \geqslant 0 \quad (\text{for } x \in S)$$

$$\lambda(x) = 0 \quad (\text{for all but a finite number of } x \text{ in } S)$$

$$\lambda_r \geqslant 0 \quad (\text{for } r = 1,2,\ldots,n)$$

It can be shown that $V_{\mathrm{I}*} = V_{\mathrm{II}}$ and that there is a solution $\lambda_1^*,\ldots,\lambda_n^*$ which is optimal for Program II. Furthermore, the following complementary slackness conditions hold:

$$\sum_{r=1}^{n} e_r^* u_r(x) = \phi(x) \quad (\text{for } x \in \Omega(\lambda^*))$$

where $\Omega(\lambda^*)$ is the support set of λ^*,

$$e_r^* = E_r \quad (\text{if } \lambda_r^* \neq 0)$$

and

$$\sum_{x \in S} \lambda^*(x) u_r(x) - \lambda_r^* = \mathrm{d}G_r(e_r^*)/\mathrm{d}e_r \quad \text{if } e_r^* \neq 0$$

We shall now present an economic interpretation of the optimal solution of Program II. $\lambda^*(x)$ can be viewed as the marginal value of a unit reduction in pollution at point x given that we have implemented the reductions $e_1^*,\ldots,e_n^*$. Thus the dual program can be used to help to make explicit the implicit values that were used to determine the standard $\Psi(\cdot)$. This could be of help in determining realistic standards that balance the value of clean air against the economic impact on an area of requiring its industry to reduce pollution. Each λ_r^* can be viewed as the marginal disutility of not permitting source r to reduce more than E_r. Thus if $\lambda_r^* > 0$ we are saying that considerations other than the simple economic criterion which we are using were used to set a limit on the reduction by source r.

Some prototype examples for Program I* when the cost function G is linear have been given by Gustafson and Kortanek (1972, 1973a), Krabs (1975), and van Honstede (1979); van Honstede (1979) has also given a nonlinear example. A test example designed to illustrate solely the climatological dispersion model is given by Brubaker et al. (1977) and Turner and Hrenko (1978). Supplemented with appropriate cost functions, this more recent test example can also be used to illustrate the optimization model set forth earlier.

3 DESIGN OF AIR-POLLUTION MONITORING NETWORKS

Generally, the effectiveness of an air-quality management program depends greatly on the ability to estimate accurately the ambient air-pollution levels throughout a given region. In turn the ability to make accurate estimates depends on the design of the monitoring network, specifically on the locations of the measuring equipment. We have developed a mathematical method for allocating pollution-measuring resources to satisfy the need for accurate estimation of the ground-level concentration of a pollutant throughout a region. The techniques are source oriented in that they give estimates of the pollution contribution of each source, which is important for use in an air-quality management program. Results from a diffusion model are used to determine the form of a response surface with which one estimates the pollutant concentration at each point in the region, including those points where measurements are not made. Multivariate regression analysis can be used to fit the response surface to the measurements obtained from a monitoring network by computing estimates of the emission distribution of point sources. Before one actually takes measurements and solves for these emission rates, one first seeks to allocate the measurement resources to points throughout the region and thus to determine the sampling sites. We have developed computational methods to allocate these resources so that best linear estimators of the unknown parameters are optimal with respect to some function of their covariance matrix in order that the regression will give good final estimates of the parameters. This is a problem in regression experimental design. For this problem design criteria are examined which tend to make the uncertainty in the estimates of the parameters as small as possible in an economically efficient manner. The assumptions needed in the experimental design are stated in Gribik et al. (1976) where the approach is illustrated using field data from Allegheny County.

Instead of providing algorithmic details we shall present the underlying statistical equation and a formal definition of the concept of experimental design.

On the basis of eqn. (1) we assume that

$$\tilde{\chi} = \sum_{r=0}^{n} \theta_r u_r + \eta \tag{2}$$

where $\tilde{\chi}$ is the actual air quality after implementation of reductions, $u_0, u_1, \ldots, u_n$ are defined as in eqn. (1), for each x $\eta(x)$ is a stochastic variable with expected value 0, and θ is a parameter vector to be estimated (see Gustafson and Kortanek, 1976).

As in the development of eqn. (1), the u_r are known, and one wants to use measurements of $\chi_e(x)$ at the receptor point x to see whether the air-quality standard is being met or not. If the selected reductions e are followed, then it should be the case that $\theta_r = 1 - e_r$ for $r = 1,2,\ldots,n$. The stochastic variable η models the situation in which neither the atmospheric-dispersion functions nor the measurements are exact. It is not known in advance whether $\chi_e(x) > \Psi(x)$ at the receptor point x, and hence χ_e must be estimated between the sampling stations, i.e. off the network.

The economic resources available for monitoring networks are always limited. Let R be a measure of the available resources for sampling involving choices between different monitoring instruments, decisions on the number of samples to be taken during the sampling period, etc.

Let x_i in the air-quality region S denote the site of a sampling station. We assume that the statistical variance in the estimate of $\chi_e(x_i)$ is of the form

$$1/R_{p_i} w(x_i) \tag{3}$$

where p_i is the fraction of available resources allocated to sampling at receptor point x_i and w is a strictly positive weighting function over all of S which is independent of any sampling procedure.

An adopted sampling program is determined by R and the table

$$\begin{matrix} x_1, x_2, \ldots, x_N \\ p_1, p_2, \ldots, p_N \end{matrix} \tag{4}$$

where x_i is a sampling point and p_i the proportion of R allocated to sampling at x_i. Thus $p_1 + p_2 + \ldots + p_N = 1$.

Gribik et al. (1976) and Gustafson and Kortanek (1976) give algorithms for determining the number N of sampling stations, the sampling sites $x_1, x_2, \ldots, x_N$, and the allocation $p_1, p_2, \ldots, p_N$ of resources between the sites. These computer methods are guided by user-oriented choices of criteria to make the maximal uncertainty of the predictions as small as possible.

Often, because of realistic economic considerations, a monitoring network must be sparse, and N is small. This may present serious difficulties in estimating the full parameter vector θ. Under these conditions it is nevertheless important to obtain reasonable estimates of air quality at points off the sparse monitoring network.

In the next section we present a method of interpolation geared to meeting the need for an objective assessment of air quality throughout the region, i.e. geared to providing improved estimates of the entire air-pollution concentration field by using actual measurements.

4 A GENERAL APPROACH TO THE PROBLEM OF THE AIR-POLLUTION CONCENTRATION FIELD

In work supported by the United States National Science Foundation and the United States Environmental Protection Agency a basis has been established for the idea that real air-quality measurements at typical locations can be combined with appropriate atmospheric-dispersion functions to provide an air-quality interpolation concentration function for an entire region (see Gustafson et al., 1977). The interpolation concentration function does not require complete information on the real source strengths of the pollutant emissions such as would be required for the application of a conventional source-oriented air-quality model. The interpolation function, however, actually involves the concept of a hypothetical or fictitious "pseudo-source" distribution, which is computed for a given atmospheric-dispersion situation or weather state by optimized fitting techniques from the measured air-quality values at selected stations and the specified atmospheric-dispersion functions. Similar ideas have recently been exploited by Heimbach and Sasaki (1975) in fitting an analytical dispersion model to sparse data on air quality. Their work was also motivated by the fact that detailed emission inventories are frequently difficult to obtain, and the predictions of source-oriented air-quality models should be capable of improvement when air-quality measurements are available.

Since the interpolation approach is self-correcting, the air-quality estimates are less affected by inaccuracies in the real emissions data. In a sense, the interpolation approach provides a technique for adjusting or "calibrating" the air-quality estimates provided by a conventional source-oriented model in order to correct for errors in the estimates of the emissions. The calibration technique is, however, not subject to the fundamental objection that has been raised in connection with the calibration procedures normally employed with source-oriented models (see Brier, 1973).

It is, in principle, the complement of another approach using empirical data recently suggested by Calder et al. (1975). In this approach multiple-station observations of air quality are coupled with estimates of the emissions for a multiple-source distribution, to deduce the effective atmospheric-dispersion functions, which can then be used as a basis for a conventional source-oriented air-quality model. The pseudo-source interpolation technique, in contrast, estimates "effective" emissions in terms of the observed air quality and prescribed atmospheric-dispersion functions. In principle, the end result is a "calibrated" interpolated air-quality model. Since the interpolation functions can be computed in advance, the technique has potential for use in air-quality prediction on a real-time basis in the presence of many weather states.

In the next section we explain how the method works.

5 THE PSEUDO-SOURCE CONCEPT AND AN APPROXIMATION METHOD FOR THE MULTIPLE-WEATHER-STATE PROBLEM

Since we shall be aggregating the original sources of pollution in certain ways, it is convenient to restate our basic equation [eqn. (1)] in a slightly different notation. The total concentration $C(x)$ at the point x will be approximated by

$$C(x) = \sum_{j=1}^{N} q_j \nu_j(x) \tag{5}$$

where q_j is the strength of source j (in, say, pounds of SO_2 per day) and ν_j is the concentration produced by a unit source at the location of source j. (A uniform background concentration B extending over the whole region is formally realized by setting $q_0 = B$ and $\nu_0(x) = 1$ for all x.) The function ν_j is of course independent of source strength and is assumed to be a computable function of the meteorological conditions such as wind direction, wind speed, atmospheric stability, and mixing depth. The functions ν_j are thus referred to as true dispersion functions. The particular combination of these conditions existing during any one steady-state interval of the time sequence is regarded as constituting one of an ensemble of possible meteorological or "weather" states. The expression of the total concentration $C(x)$ as a linear superposition extending over the atmospheric-dispersion functions $\nu_j(x)$ is mathematically analogous to the use of polynomial or trigonometric functions for the finite-expansion approximation of continuous functions.

An aggregated pseudo-source formulation shall be in terms of a much smaller number of sources n than the number N of real identified sources employed in eqn. (5). Formally, we proceed as follows. Let $j_0, j_1, \ldots, j_n$ be integers such that

$$0 = j_0 < j_1 < \ldots < j_n = N$$

We then combine sources with indices between $j_0 + 1 (= 1)$ and j_1 into the first pseudo-source, those with indices from $j_1 + 1$ to j_2 into the second pseudo-source, and so on.

Generally, the pseudo-source r consists of sources with indices from $j_{r-1} + 1$ to j_r. Let the concentration contribution from the rth pseudo-source be U_r. Thus eqn. (5) becomes

$$C(x) = \sum_{r=1}^{n} U_r(x) \tag{6}$$

where

$$U_r(x) = \sum_{j=j_{r-1}+1}^{j_r} q_j v_j(x)$$

We now write $U_r(x)$ as

$$U_r(x) = Q_r u_r(x)$$

where

$$Q_r = \sum_{j_{r-1}+1}^{j_r} q_j \qquad u_r(x) = U_r(x)/Q_r \tag{7}$$

Q_r is the combined strength of all the sources in source class r. The functions $u_r(x)$ are still dependent on source-strength considerations since they depend on the ratio of the individual source strengths in the class to the class total. This is in contrast to the functions $v_j(x)$ of eqn. (5) which are true dispersion functions that are entirely independent of source-strength considerations. The pseudo-source concept is only entirely independent of knowledge of the real source distribution in the obvious case $n = N, j_r = r$, when $Q_r = q_r$, so that $u_r(x) = v_r(x)$. However, in this case the determination of the large number of values of q_j ($j = 1,2,...,N$) in eqn. (5) relative to measurement data from a sparse sampling network is essentially impossible. One of the primary reasons for the pseudo-source method is to reduce the number of sources to be commensurate with sparse measurement data.

5.1 The Aggregated–Averaged Approximation Method

Another case is very important as an approximation. It is obvious that the real sources may be aggregated into classes in such a manner that within each class the strengths are "approximately" equal, i.e. equal to within some prescribed variation. This motivates an important specialization of the aggregated-source concept in which every source strength in a class is replaced by the average value, say $\overline{q}(r)$, for the class. This approximation can be made as close as we wish up to the limit of the completely unaggregated situation. It will be designated as the Aggregated–Averaged Approximation. For this case we can see from eqn. (6) that

$$U_r(x) \simeq \overline{q}(r) V_r(x) \qquad V_r(x) = \sum_{r\ \text{class}} v_j(x) \tag{8}$$

and the function $V_r(x)$ is now completely independent of source-strength considerations. Like the original functions $v_j(x)$ it depends only on positional parameters for the sources constituting the class and on the meteorological conditions affecting dispersion. Evidently from eqn. (7) the same will be true for the functions

$$u_r(x) = V_r(x)/(\text{number in } r \text{ class})$$

In terms of the aggregated pseudo-source classes eqn. (5) is now replaced by a sum that only involves n terms:

$$C(x) = \sum_{r=1}^{n} Q_r u_r(x) \tag{9}$$

If concentrations are measured at p sampling stations $x_1, x_2, \ldots, x_p$ yielding values C_i, $i = 1, \ldots, p$, we have the following linear system of equations to determine the n pseudo-source strengths Q_r $(r = 1, 2, \ldots, n)$:

$$\sum_{r=1}^{n} Q_r u_r(x_i) = C_i \qquad (i = 1, 2, \ldots, p) \tag{10}$$

Generally, eqn. (10) may have to be solved in the least-squares sense, i.e. the sense for which

$$\Delta = \min_{Q} \sum_{i=1}^{p} \left[\sum_{r=1}^{n} Q_r u_r(x_i) - C_i \right]^2 \tag{11}$$

where the vector $Q = (Q_1, Q_2, \ldots, Q_n)$ is such as to minimize Δ (see Dahlquist and Björk, 1974). If Q_r denotes the appropriate least-squares solution in general, then on substituting back in eqn. (9) we obtain the corresponding air-quality interpolation formula for the entire concentration field based on the use of pseudo-sources:

$$\hat{C}(x) = \sum_{r=1}^{n} \hat{Q}_r u_r(x) \tag{12}$$

This is a solution in the least-squares sense of eqn. (11); i.e.

$$\sum_{i=1}^{p} [\hat{C}(x_i) - C_i]^2 = \text{minimum}$$

The notation "^" is used here, since in general it will be necessary to consider inconsistent sets of eqns. (10) having no exact mathematical solutions but only approximate solutions as, for example, in a least-squares sense. In this case of course the values of $\hat{C}(x_i)$ for the sampling locations x_i will not exactly duplicate the measured values C_i $(i = 1, 2, \ldots, p)$. Finally we note that in principle the dispersion functions $u_r(x)$ could be precalculated, once and for all, for a grid of points. The interpolated concentration distribution could then be very simply calculated from eqn. (12) in terms of the $\hat{Q}_r$, which from eqn. (10) are functions only of the observed values C_i and $u_r(x_i)$ $(i = 1, 2, \ldots, p)$. This offers the possibility of interpolation on a real-time basis.

5.2 Application to the Multiple-Weather-State Problem

The basic idea is that similar air-pollution concentration patterns should be related to similar meteorological patterns. We now indicate how we intend to treat the large number of meteorological conditions that influence the transport of pollutants.

Let Ω be the set of all weather states w, each as specified by, say, wind direction, wind speed, stability class, and mixing depth. For example, one might specify 16 wind directions, ten wind-speed classes, six stability classes, and four different mixing depths (see Calder et al., 1975). Ω then contains $16 \times 10 \times 6 \times 4 = 3840$ weather states. For each of these we shall have an equation of the form (12):

$$C^w(x) = \sum_{r=1}^{n} Q_r u_r^w(x) \tag{13}$$

In principle we could now proceed using least-squares analysis and for each weather state could determine the appropriate interpolation function $m(x)$ as a basis for the general interpolation formula $\hat{C}^w(x)$ for the concentration at an arbitrary location x for the weather state w occurring. However, in view of the large number of weather states and the complexity of their actual fields this could involve a prohibitive amount of computation. We therefore consider a partitioning of the set Ω of all weather states into a smaller number of subsets or weather classes

$$\Omega_1, \Omega_2, \ldots, \Omega_j$$

and examine the question of whether an adequate general concentration interpolation formula can be derived for $\hat{C}^w(x)$ in terms of interpolation functions $m(x)$ calculated for each of the smaller number of classes, rather than for the large number of individual states.

Since partitioning is a combinatorial problem it can be accomplished in many ways, and further study is required to compare the various possibilities. Only a very limited initial study is presented in Gustafson et al. (1977). However, for each weather class Ω_k of a partitioning we shall need to examine whether a single interpolation function $m(x)$ can be employed (to avoid confusion we refrain from using the index symbol k on $m(x)$) such that in matrix notation

$$A^{\mathrm{T}} m(x) = u^w(x)$$

or

$$\sum_{i=1}^{p} u_r^w(x_i) m_i(x) = u_r^w(x) \qquad (r = 1,2,\ldots,n;\ \text{all } w \text{ in } \Omega_k) \tag{14}$$

and

$$\hat{C}^w(x) = m^{\mathrm{T}}(x) C^w = \sum_{i=1}^{p} m_i(x) C_i^w \qquad (\text{all } w \text{ in } \Omega_k) \tag{15}$$

Here C_i^w denotes the measured concentration value at the sampling station i $(i = 1,2,\ldots,p)$ for the weather state w that is occurring. It may be noted that eqn. (15) follows immediately from eqn. (14) on using the pseudo-source strengths $\hat{Q}_r$ defined by the solution of eqn. (13), for

$$\hat{C}^w(x) = \sum_{r=1}^{n} \hat{Q}_r u_r^w(x) = \sum_{r=1}^{n} \hat{Q}_r \left[\sum_{i=1}^{p} u_r^w(x_i) m_i(x) \right]$$

$$= \sum_{i=1}^{p} m_i(x) \left[\sum_{r=1}^{n} \hat{Q}_r u_r^w(x_i) \right]$$

$$= \sum_{i=1}^{p} m_i(x) C_i^w$$

The basic concept of the pseudo-source technique for the multiple-weather-state situation is thus to determine the interpolation functions $m_i(x)$ by least-squares solution (if necessary) of the system (14) and then to substitute into eqn. (15) to obtain the required interpolation formula. As we have noted, the method can be applied irrespective of whether the system is overdetermined, evenly determined, or underdetermined. Our preliminary and tentative conclusion is that the interpolation functions should be entirely independent of source-strength considerations and only dependent on the source positional parameters and the weather class. Because of the approximations that have been noted this conclusion may only be valid in an approximate sense, and the degree of approximation must be further tested experimentally.

5.3 Data Management and Scenario Testing of the Methodology

An example of a practical scenario for regulating sulfur fuels could involve a combination of requirements on the sulfur content of fuels, requirements on the desulfurization of oil, or requirements on the purification of flue gases. For example, in the production of steel and iron a scenario usually consists of various practical combinations of the following control methods: (1) allocation levels of natural gas to boilers, (2) sulfur levels of fuels, (3) fuel mixes (coal, natural gas, fuel oils), and (4) levels of coke-oven gas desulfurizations. In addition we have studied scenarios related to air-pollution episodes such as those associated with second-stage air-pollution alerts. In principle the specification of scenarios will be aided by the optimization models studied in Section 2.

Formally, a scenario is a specification of a particular emission distribution $\overline{q}_1, \overline{q}_2, \ldots, \overline{q}_N$ which determines a total concentration $\overline{C}(x)$ via eqn. (5). With the specification of a scenario a whole family of isopleths is generated for the region. However, these families can actually be rigorously defined for weather states w. This gives rise to $\overline{C}^w$ [see eqn. (13)] where now $\overline{Q}_1, \ldots, \overline{Q}_n$ stem from a particular specification of a pseudo-class selection and weather state. Therefore, under the particular fixed scenario, the value $\overline{C}^w(x_i)$ corresponds to a measurement at the sampling station at x_i during a weather state w.

Thus we have a way of testing the worth of methods of forming pseudo-source classes and weather-state partitions. For any pseudo-source class and weather-state partitioning we solve eqn. (14) for the interpolating functions m_i, one for each sampling station. We then compute

$$\widetilde{C}^w(x) = \sum_{i=1}^{p} m_i(x)\overline{C}^w(x_i)$$

and claim that under a good pseudo-source classification and weather-state partition it should be the case that $\widetilde{C}^w(x)$ values are close to $\overline{C}^w(x)$ values.

If a good fit is realizable over a variety of scenarios, then a sound basis is established for the application of the interpolation functions m_i determined in eqn. (14) to actual measurement data at the sampling stations. This validation procedure could apply to any air-quality region over which an atmospheric-dispersion function was applicable.

In a region where there is a substantial terrain effect we conjecture that the region can be decomposed into individual areas, termed hot spots, within which concentration peaks are attained under a variety of meteorological conditions. In this approach the aggregation assumption relating to sources that are not in a particular hot spot is that SO_2 concentrations transported from sources outside the region are uniform. In this sense the many sources outside a particular hot spot are "aggregated".

6 CONCLUSIONS

Over the last ten years there have been many studies justifying the conclusion that a least-cost optimization can achieve compliance with air-quality standards at a lower cost than can a proportional emissions-reduction policy. The conclusion is certainly not surprising and is the same whether the model is of the "continuous" variety or one of the discrete integer-programming types.

Virtually all these optimization studies, continuous or discrete, begin with a preselected grid of receptor points. The surprising conclusion of our work is that it is not necessary to make this choice with our model, for our optimization approach employs the complete air-pollution concentration field as it is currently known. The optimization process itself, namely the problem to be solved, determines the relevant, generally nonuniform, grid of points where concentrations are at their permitted maximum air-quality standard values. The key ingredient that distinguishes our optimization model from all others that we have seen is the way in which we formulate the phrase "throughout the region" in the legislative mandate "achieve . . . ambient air-quality standards throughout the region".

Recognizing that the air-pollution concentration field is one of great complexity, we have developed a new method of combining real air-quality measurements taken at typical locations in the region with atmospheric-dispersion functions to provide an air-quality interpolation concentration function for the entire region. This function is computed from highly disaggregated information involving many diverse atmospheric weather states together with "pure" source-oriented dispersion functions which do not depend on the emission distribution of the sources themselves. The method is designed to operate on a real-time basis and is self-correcting. One of its key strengths is that it applies to regions where the air-quality monitoring network is sparse and where information on the real source strengths is generally unavailable or unreliable.

Finally we have reviewed a statistical optimization method for the problem of locating air-monitoring instrumentation. The approach rests heavily on classical and well-known work in the theory of optimal experimental regression design. It thereby provides a sound and dependable statistical basis for solving the problem.

The statistical methods in this paper are a complement to the selective abatement-optimization procedures introduced in the earlier sections. Scenarios which result from these optimization studies may now be verified in the field, including the case where because of economic realities the air-quality monitoring network is necessarily sparse.

ACKNOWLEDGMENTS

The authors are indebted to many individuals for key discussions on the topics in this paper. Those to whom we owe our greatest debt include K.L. Calder, R. Carbone, R. Dunlap, W.L. Gorr, P.R. Gribik, S. Hanna, D.N. Lee, J. Mahoney, L. Olsson, E.S. Rubin, J. Seinfeld, J. Sweigart, and D.B. Turner.

The work of one of the authors (K.O. Kortanek) was partially supported by National Science Foundation Grant ENG 78-25488.

REFERENCES

Begränsning av svavelutsläpp – en studie av styrmedel (1974). SOU, 101 Allmänna Förlaget, Stockholm, Sweden. (Summary in English.)

Brier, G.W. (1973). Validity of the air quality display model calibration procedure. EPA Rep. EPA-R4-73-017. Environmental Protection Agency.

Brubaker, K.L., Brown, P., and Cirillo, R.R. (1977). Addendum to user's guide for climatological dispersion model, EPA-450/3-44-015. Environmental Protection Agency.

Calder, K.L. (1977). Multiple-source plume models for urban air pollution – their general structure. Atmos. Environ., 11:403–414.

Calder, K.L., Breiman, L., and Meisel, W.S. (1975). Empirical techniques for analyzing air quality and meteorological data: Part II. Feasibility of a source-oriented empirical air quality simulation model. Rep. ESRL-RTP-054, Ser. No. 4. Environmental Protection Agency.

Carbone, R. and Sweigart, J.R. (1976). Equity and selective pollution abatement procedures. Management Sci., 23:361–370.

Carbone, R., Gorr, W.L., Gustafson, S.-Å., Kortanek, K.O., and Sweigart, J.R. (1978). A bargaining resolution of the efficiency versus equity conflict in energy and air pollution regulation. TIMS Studies in Management Studies, 10:95–108.

Dahlquist, G. and Björck, A. (1974). Numerical Methods. Prentice–Hall, Englewood Cliffs, New Jersey.

Fahlander, J., Gustafson, S.-Å., and Olsson, L.E. (1974). Computing optimal air pollution abatement strategies – some numerical experiments on field data. In Proc. Meet. NATO–CCMS Panel on Air Pollution Modeling, 3rd. Environmental Protection Agency, Raleigh, North Carolina, Chap. 27.

Gorr, W.L. and Kortanek, K.O. (1972). Numerical aspects of pollution abatement problems: optimal control strategies for air quality standards. In M. Henk, A. Jaeger, R. Wartman, and H.K. Zimmerman (Editors) Proc. in Operations Research. Physica, Wurzburg, FRG.

Gorr, W.L., Gustafson, S.-Å., and Kortanek, K.O. (1972). Optimal control strategies for air quality standards and regulatory policy. Environ. Plan., 4:183–192.

Gribik, P.R. (1978). Selected applications of semi-infinite programming. In C.V. Coffman and G.J. Fix (Editors) Constructive Approaches to Mathematical Models: a Symposium in Honor of R.J. Duffin. Academic Press, New York, to be published.

Gribik, P.R., Sweigart, J.R., and Kortanek, K.O. (1976). Designing a regional air pollution monitoring network: an appraisal of a regression experimental design approach. In Proc. EPA Conf. on Environmental Modeling and Simulation.

Gustafson, S.-Å., and Kortanek, K.O. (1972). Analytical properties of some multiple-source urban diffusion models. Environ. Plan., 4:31–43.

Gustafson, S.-Å. and Kortanek, K.O. (1973a). Mathematical models for air pollution control: numerical determination of optimizing abatement policies. In R.A. Deininger (Editor) Models for Environmental Pollution Control. Ann Arbor Science Publishers, Ann Arbor, Michigan.

Gustafson, S.-Å. and Kortanek, K.O. (1973b). Mathematical models for optimizing air pollution abatement policies: numerical treatment. In J. Smoli (Editor) Proc. Bilateral US–Czechoslovakia Environmental Protection Semin., 3rd. Dum Techniky, CVTS, Prague.

Gustafson, S.-Å. and Kortanek, K.O. (1973c). Sensitivity aspects, on mathematical programs for air pollution abatement policies. In E. Shlifer (Editor) Proc. TIMS Conf. on Management Sciences, Developing Countries, and National Priorities, 20th. Jerusalem Academic Press, Jerusalem.

Gustafson, S.-Å. and Kortanek, K.O. (1974). Determining sampling equipment locations by regression experimental design with application to environmental pollution control and acoustics. In Proc. Computer Science and Statistics Annu. Symp. on the Interface, 7th. Iowa State University, Ames, Iowa.

Gustafson, S.-Å. and Kortanek, K.O. (1975). On the calculation of optimal long-term air pollution abatement strategies for multiple-sources areas. In Proc. Int. Conf. on Mathematical Models for Environmental Problems, University of Southampton.

Gustafson, S.-Å. and Kortanek, K.O. (1976). Computation of optimal experimental designs for air quality surveillance. In P.-J. Jansen, O. Moeschlin, and O. Rentz (Editors) Quantitative Modelle für Ökonomisch–Ökologische Analysen. Anton Hain, Misenheim.

Gustafson, S.-Å., Kortanek, K.O., and Sweigart, J.R. (1977). Numerical optimization techniques in air quality modeling: objective interpolation formulas for the spatial distribution of pollutant concentration. J. Appl. Meteor., 16:1243–1255.

Heimbach, J.A. and Sasaki, Y. (1975). A variational technique for mesoscale objective analysis of air pollution. J. Appl. Meteor., 14:194–203.

van Honstede, W. (1979). An approximation method for semi-infinite problems. In R. Hettich (Editor) Semi-infinite Programming. Springer, Berlin.

Hrenko, J.M. and Turner, D.B. (1975). An efficient gaussian-plume multiple-source air quality algorithm. Annu. Meet., APCA, Boston, 68th. Paper 75-04.3.

Krabs, W. (1975). Optimization and Approximation, B.G. Teubner, Stuttgart, FRG.

Larsen, R.I. (1969). A new mathematical model of air pollutant concentration averaging time and frequency. J. Air Poll. Contr. Assoc., 19:24–30.

Turner, D.B. and Hrenko, J. (1978). User's Guide for RAM, Volume 1. Algorithm Description and Use. EPA-600/8-78-0/6a. Environmental Protection Agency.

SOURCE ALLOCATION AND DESIGN VIA SIMULATION MODELS

H.G. Fortak
Freie Universität, Berlin (West)

1 INTRODUCTION

Until recently, the problem of climatological air-pollution forecasting for air-pollution abatement arose either when new emission sources (with known characteristics) were planned or when improvements to already existing sources were attempted. Now both problems can be solved by means of mathematical–meteorological simulation models (Fortak, 1972).

The problem becomes far more complicated if certain limits for the air-pollution climatology of a region are prescribed and if one requires the feasible installation of industrial complexes of which the total source emission and source configuration do not exceed the prescribed concentrations. In the FRG this question arose for the first time during the planning phase of the Neuwerk–Scharhörn deep-sea harbor.

The solution of problems of air-quality management planning does not lie in monitoring and controlling individual concentrations, at a few locations within the area considered, produced by the emission sources. Because of the high variability of meteorological conditions for transport and diffusion a broad spectrum of concentrations is produced at each location. Therefore it is necessary to derive frequency distributions of air-pollution concentrations from observations over a sufficiently long period of time. The frequency distributions are characterized by a set of statistical parameters which form the basis of what has been called "air-pollution climatology" (Fortak, 1972). These parameters are, for instance, the mean value of the concentrations for a given period of time and the so-called percentiles, i.e. the percentage of observations that show concentrations exceeding a certain limit.

When trying to solve environmental *planning* problems, it is evident that *observed* concentrations are not available. Therefore the development of concentrations over time has to be simulated by means of air-pollution models in order to derive the air-pollution climatology. For this purpose a certain "normal development" of meteorological conditions for transport and diffusion is derived from long-term meteorological observations in the region considered. This normal development is then divided into classes of actual weather situations, and for each a concentration field is calculated. The knowledge of the frequency distributions of these weather situations (conditions for transport and diffusion) allows the derivation of frequency distributions of pollutant concentrations for any location within the area. This method of simulating an air-pollution climatology was first applied to the city of Bremen (Fortak, 1966, 1970) and will be applied to the present case also, for until now it has been the only reasonable method for dealing with air-pollution climatology forecasts.

In detail, the method is as follows. First statistics of the meteorological conditions for transport and diffusion for the flats and coastal area are obtained. Since synoptic observations for Scharhörn are not available, wind observations for Scharhörn are related to synoptic observations for Nordholz. A total source strength is then assumed for the deep-sea harbor area and this is uniformly distributed over various source configurations. With regard to the frequency distribution of the meteorological conditions for transport and diffusion, the characteristic air-pollution climatology parameters are simulated for all source configurations and particularly for a wind direction of 300° (i.e. the direction from the deep-sea harbor via Neuwerk to Sahlenburg–Duhnen), which is the wind direction with the most interesting effects. The field representation of the parameters for the whole region completes this investigation.

2 BASIC CHARACTERISTICS OF THE APPLIED SIMULATION MODEL

Because of the large number of cases to be calculated it was only possible to adopt the simplest but still the most appropriate model (Pasquill, 1962; Fortak, 1970; Fett, 1974). Here the spatial distribution of concentrations produced by a single source of strength Q(mg/s) is given by

$$S(x,y,z) = \frac{Q}{\bar{U}_1} \frac{\exp(-y^2/2\sigma_y^2)}{\sqrt{2\pi}\sigma_y} \left[\frac{\exp\{-(H_{\text{eff}} - z)^2/2\sigma_z^2\}}{\sqrt{2\pi}\sigma_z} + \frac{\exp\{-(H_{\text{eff}} + z)^2/2\sigma_z^2\}}{\sqrt{2\pi}\sigma_z} \right] \tag{1}$$

assuming unlimited vertical dispersion ($S \to 0$ if $z \to \infty$) and reflection of the pollutant at the Earth's surface ($\partial S/\partial z = 0$ if $z = 0$), i.e. a non-absorbing ground.

The assumption of unlimited vertical dispersion can be removed by introducing a "lid" caused by a temperature-inversion layer (Fortak, 1961, 1970) but this was omitted since these weather situations are rare in the region considered. The partial absorption of pollutants at water surfaces was included and was estimated by a series of experimental calculations. The boundary condition at the Earth's surface is then replaced by

$$K_z \frac{\partial S}{\partial z} = \beta S; \qquad z = 0$$

which expresses the fact that a fraction of the turbulent vertical transport $K_z \partial S/\partial z$ of pollutants, proportional to the ground-level concentration S, is absorbed. This part of the investigation is feasible only by means of a very complex numerical method (Fortak,

1964) and is not worthwhile given the incomplete knowledge of the absorption parameter β. Therefore it was only possible to estimate the effect of absorption at water surfaces up to the limit of total absorption. This can be formulated simply by eqn. (1) where the sign between the two terms in parentheses becomes negative. Denoting the solution of eqn. (1) by S_{refl} for the assumption of a non-absorbing ground and by S_{abs} in the case of total absorption, the difference between the two limiting cases is

$$\Delta S = S_{refl} - S_{abs} = 2\frac{Q}{\bar{U}_1}\frac{\exp(-y^2/2\sigma_y^2)}{\sqrt{2\pi}\sigma_y}\frac{\exp\{-(H_{eff}+z)^2/2\sigma_z^2\}}{\sqrt{2\pi}\sigma_z} \quad (2)$$

$S_{abs} = 0$ at the Earth's surface in the case of total absorption, i.e. $\Delta S = S_{refl}$ if $z = 0$. For all layers $z > 0$, a finite value of S_{abs} is obtained and can be compared with S_{refl}. On introducing

$$\sigma_y = Fx^f; \qquad \sigma_z = Gx^g \quad (3)$$

for the standard deviations, the maximum downwind values ($y = 0$) of ΔS are to be found at a distance

$$x_{max}(\Delta S) = \left(\frac{H_{eff}+z}{G\sqrt{2r}}\right)^{1/g} \qquad \left(\text{where } 2r = \frac{f+g}{g}\right) \quad (4)$$

from the emission source and are given by

$$(\Delta S)_{max} = 2\frac{Q}{\bar{U}_1}\frac{1}{\sqrt{2\pi}\sigma_y(x_{max})}\frac{\exp(-r)}{\sqrt{2\pi}\sigma_z(x_{max})} \quad (5)$$

Since eqn. (4) with $z = 0$ and eqn. (5) together with eqn. (4) represent exactly the maximum distance $x_{max}(S_{refl})$ and the maximum ground-level concentration $(S_{refl})_{max}$, the effect of absorption is highest in the neighborhood of maximum ground-level concentration values. This is particularly true when the concentration layer z is low compared to the effective stack height H_{eff}.

By deriving the expression for the turbulent downward flux of pollutants at the Earth's surface in the case of total absorption, $K_z \partial S_{abs}/\partial z$ for $z = 0$, and by integrating this expression successively over the entire domain in the crosswind direction ($-\infty < y < \infty$) and from the emission source ($x = 0$) to a certain distance x, respectively, we obtain the total amount of pollutants being absorbed up to a distance x from the emission source

$$A(x) = \int_0^x dx \int_{-\infty}^{\infty} K_z(\partial S_{abs}/\partial z)\, dy = Q\, \text{erfc}\left\{\frac{H_{eff}}{\sqrt{2}\,\sigma_z(x)}\right\} \quad (6)$$

where erfc $X = 1 - \text{erf}\, X$ and erf $X = (2/\sqrt{\pi})\int_0^X \exp(-t^2)\, dt$ is the well-known error integral.

Analysis of eqn. (6) shows that, for all meteorological conditions, in the case of total absorption at water surfaces a considerable amount of pollutants is filtered out from the atmosphere along the distance between the deep-sea harbor and the coast. Partial absorption lies between this extreme and the case of complete absence of absorption. It cannot be examined, however, without comprehensive knowledge of the absorption coefficient β.

With respect to planning problems it therefore seems to be more feasible to assume non-absorbing water surfaces. For this case the ground-level concentration is obtained from eqn. (1)

$$S(x,y,0) = 2\frac{Q}{\bar{U}_1}\frac{\exp(-y^2/2\sigma_y^2)}{\sqrt{2\pi}\,\sigma_y}\frac{\exp(-H_{\mathrm{eff}}^2/2\sigma_z^2)}{\sqrt{2\pi}\,\sigma_z} \tag{7}$$

where the mean transport velocity $\bar{U}_1$ and the effective stack height H_{eff} need to be explained.

The theory of turbulent diffusion takes into account the fact that the wind velocity increases with increasing height. The exponential law

$$U(z) = U(z_{\mathrm{A}})(z/z_{\mathrm{A}})^m \tag{8}$$

allows the estimation of the vertical velocity distribution from the measured wind velocity $U(z_{\mathrm{A}})$ at the anemometer level z_{A}. The representative transport velocity of pollutants can then be defined by taking the mean value of eqn. (8) over the polluted layer

$$\bar{U}_1 = (1/2H_{\mathrm{eff}})\int_0^{2H_{\mathrm{eff}}} U(z)\,\mathrm{d}z \tag{9}$$

where the effective stack height H_{eff} is the sum of the emission height h and the plume rise Δh according to

$$H_{\mathrm{eff}} = h + \Delta h \tag{10}$$

To date, the problem of specifying the plume rise has not been sufficiently well solved. The present investigation applies an approved representation (Fortak, 1969) in the following form

$$\Delta h = 135(E^*/\bar{U}_2^3)^{0.4} \tag{11}$$

where E^* is a technical emission value and $\bar{U}_2$ is the mean value of $U(z)$ taken over the range of the plume rise

$$\bar{U}_2 = (1/\Delta h)\int_h^{H_{\mathrm{eff}}} U(z)\,\mathrm{d}z \tag{12}$$

The basic characteristics of the applied model are now fully described. Only the standard-deviation parameters have still to be determined. They depend on the stability classes obtained from synoptic observations. For the purposes of this investigation the problem was considered anew and a more suitable classification of the parameters was found, by which certain discrepancies in the statistical results are avoided. Very comprehensive numerical investigations resulted in a modified version of the Brookhaven diffusion parameters for effective stack heights $H_{eff} \geqslant 100$ m (see Table 1) (Singer and Smith, 1966).

TABLE 1 Diffusion parameters and wind profile versus stability class.

Stability class		Diffusion parameters			Wind profile
Turner (1964)	Brookhaven	$f = g$	F	G	m
1	B_2	0.91	0.40	0.41	0.111
2	B_1	0.86	0.36	0.33	0.208
3		0.81	0.34	0.27	0.250
4	C	0.78	0.32	0.22	0.288
5		0.76	0.315	0.18	0.325
6		0.74	0.31	0.13	0.355
7	D	0.71	0.31	0.06	0.400

3 STATISTICS OF METEOROLOGICAL CONDITIONS FOR TRANSPORT AND DIFFUSION

According to Turner's method (Turner, 1964) for the determination of stability classes, not only the wind velocity but also a number of synoptic observations must be used. These data are needed at a representative location within the area of diffusion. At Scharhörn, excellent observations of wind velocity and wind direction were available (Antfang, 1969), but no other synoptic data were reported. Therefore a station close to the coast (Nordholz) was chosen, but its wind statistics were completely different from those on the island of Scharhörn. There is, however, good reason to assume that all the other synoptic data connected with the large-scale weather situation are identical on the island of Scharhörn and at Nordholz. The horizontal distance between the two locations is approximately 20 km. The only problem is to relate the wind observations at the Nordholz station to those on the island of Scharhörn for all the observation dates. A very complicated procedure was developed such that the Nordholz wind data could be modified so as to become identical with the statistics obtained by observations at Scharhörn. A typical example for the region of the Deutsche Bucht is given in Figure 1, which shows that the most frequent weather situation during July is characterized by a wind direction of 300° (the direction from the planned deep-sea harbor towards the beaches near Cuxhaven). The corresponding wind velocity is around 10 m/s. The complete statistics of the meteorological conditions for transport and diffusion were obtained in the same manner. The period of observations which forms the basis of these statistics was from April 17, 1963, to December 31, 1967. Briefly, from these statistics it follows that the region

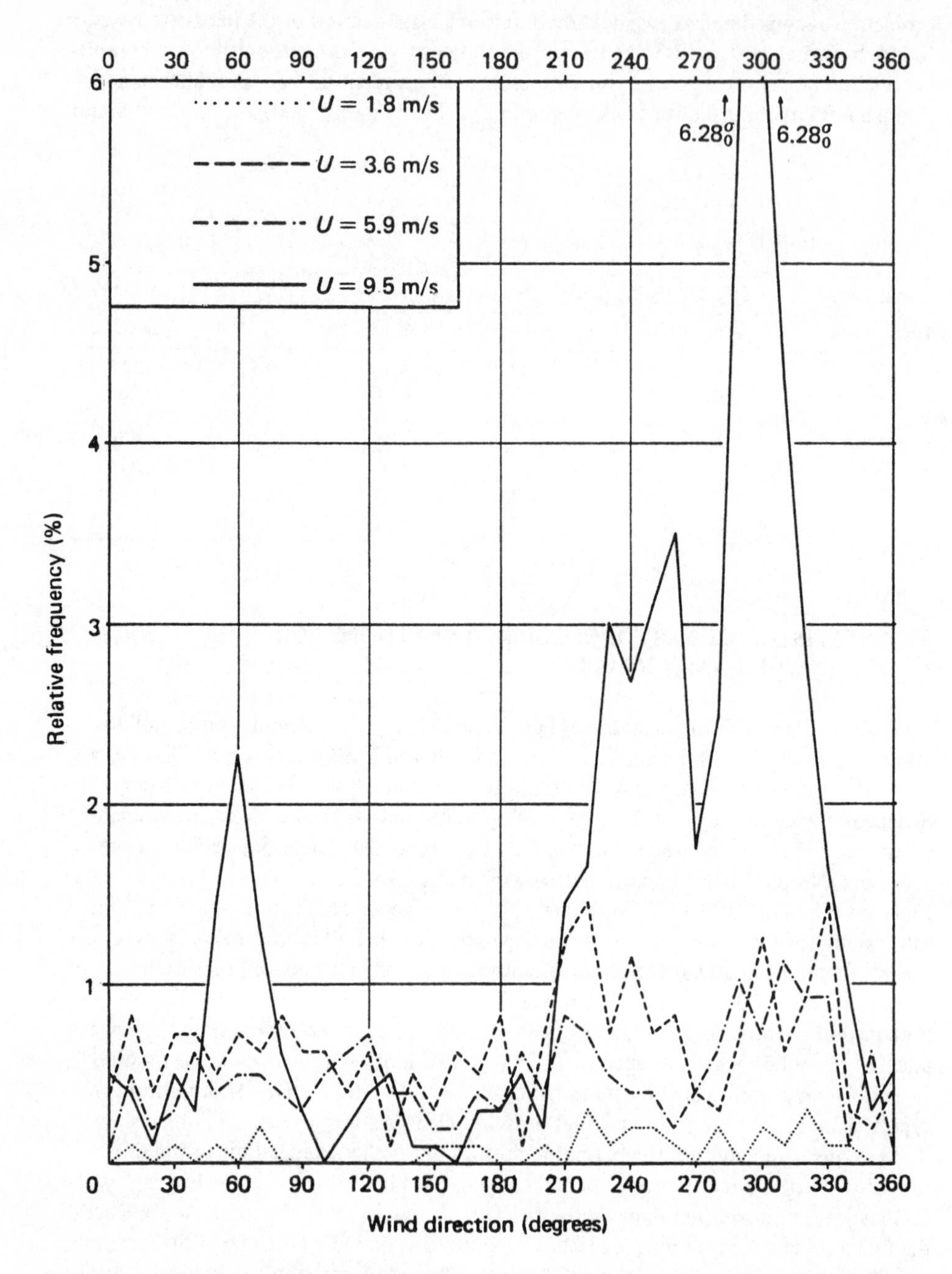

FIGURE 1 The relative frequency distribution of the wind direction and classified wind velocities for one selected month (July).

between the island of Scharhörn and the beaches near Cuxhaven is very frequently loaded with air pollution, particularly during the summer months. However, this pollution occurs in most cases with high wind velocities.

4 PROBLEMS OF SOURCE CONFIGURATION

It was difficult to predict realistic source configurations without knowing the planned structure of industrialization in the area of the deep-sea harbor. It is logical to use the stack height as an initial variable. The classification $h = 50, 100, 150, 200, 250, 300$ m was chosen. The 50-m stack height was only used for comparison. Since weather situations with fog and low-level inversions are quite frequent (Antfang, 1969), the stacks should tend to be high. Emissions will then take place above all low-level inversions and hence no ground-level concentration will be produced at any point.

This requirement for high stacks excludes other favorable solutions which assume a great number of low-level sources. With regard to the geometrical arrangement of more than one source, it is logical to arrange the sources so that they are parallel to the receptors which require the most protection. Otherwise the superposition of sources will lead to a considerable increase in ground-level concentrations. A number of numerical experiments demonstrated this superposition impressively. Finally, only five geometrical configurations were utilized: (1) a single source; (2) two sources 4 km apart; (3) five sources in line, each 1 km apart; (4) two groups of nine sources with a grid distance of 600 m; (5) a uniform distribution of 36 sources over an area of 3 km × 3 km with a grid distance of 600 m.

For each geometrical source configuration the whole set of stack heights was used to produce air-pollution climatologies for that region. In total, 30 air-pollution climatologies were actually calculated. The total source strength of the industrialized area was assumed to be 10 tonne/h for all types of air pollutant (SO_2 or NO_x). This amount was distributed equally between the sources of each configuration. Difficulties arose when we attempted to calculate the plume rise. The technical emission value depends on the type of installation, i.e. on the stack height and the source strength. Without knowledge of the actual data only statistical relationships could be used. Utilizing the source inventories of the cities of Bremen, Düsseldorf, and Frankfurt, the relationships

$$h = A\sqrt{Q}; \qquad E^* = BQ; \qquad E^* = Ch^2$$

were obtained. The factors A, B, C were obtained by regression analysis. In the case of a single source $E^* = 150Q = 1500$ (m^4/s^3) was used, but for all other source configurations $E^* = 200Q$ gave the best results.

5 RESULTS OF THE NUMERICAL EXPERIMENTS

The program was composed of two main parts. First, we calculated the parameters for the air-pollution climatology only for the wind direction 300°, which was considered to be the critical direction. For this wind direction, the frequency distributions of wind

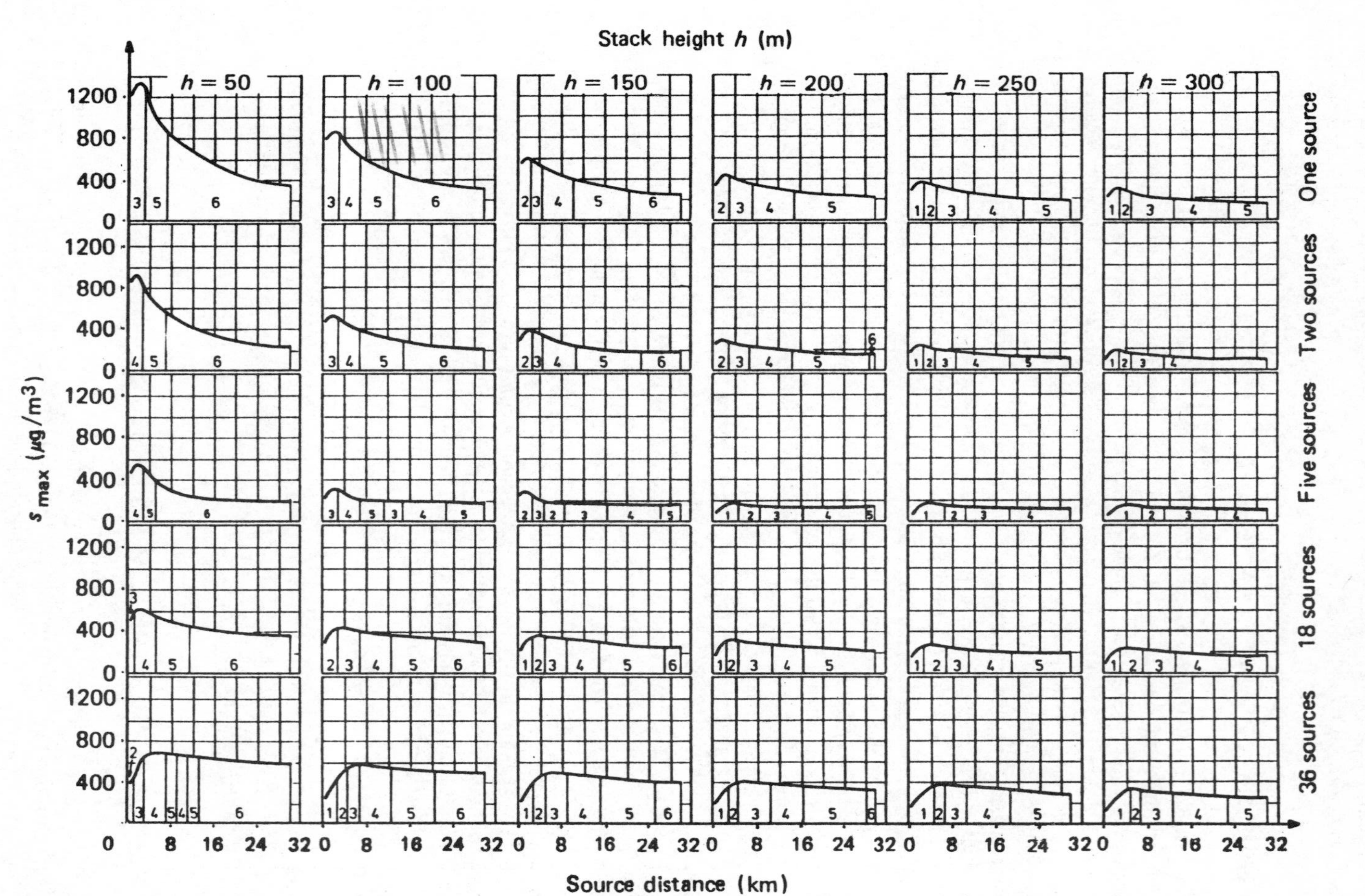

FIGURE 2 Maximum ground-level concentration of SO_2 versus source distance and stability class (for various stack heights h and source configurations).

velocity and stability classes were used, and very detailed information was obtained. The second main part of the program was concerned with calculations of two-dimensional fields of the parameters of the air-pollution climatologies. The area considered is shown in the subsequent figures.

The first pre-investigation was concerned with the maximum ground-level concentrations along the wind direction 300°. All combinations of source configuration, wind velocity, and stability class were used. Figure 2 shows the result. As an example we can consider the experiment with $h = 150$ m and five sources: as a result of the different combinations of wind velocity and stability class, two regions are produced where absolute maxima are caused by a certain stability class (2 and 3). Up to a distance of 16 km these stability classes are responsible for the highest maxima. Further downwind the maxima are less than 200 $\mu g/m^3$ and are produced by stability class 4 which occurs very frequently. From Figure 2 it is reasonable to choose first of all a source configuration which produces maximum ground-level concentrations that are as small as possible. Accordingly, Figure 2 recommends as the best solution an arrangement of five sources each 300 m high. In this case the maximum ground-level concentrations rarely exceed 100 $\mu g/m^3$. More detailed information can be obtained from Figure 3. Here the stability class is not indicated but the possible maximum ground-level concentrations are drawn for all the assumed stack heights and all source distances up to 24 km. It should be noted that the scale of the ordinate is logarithmic. It can easily be seen that the configuration with five sources is certainly the most favorable. This is particularly true for the region approximately 4 km from the source where the island of Neuwerk is located.

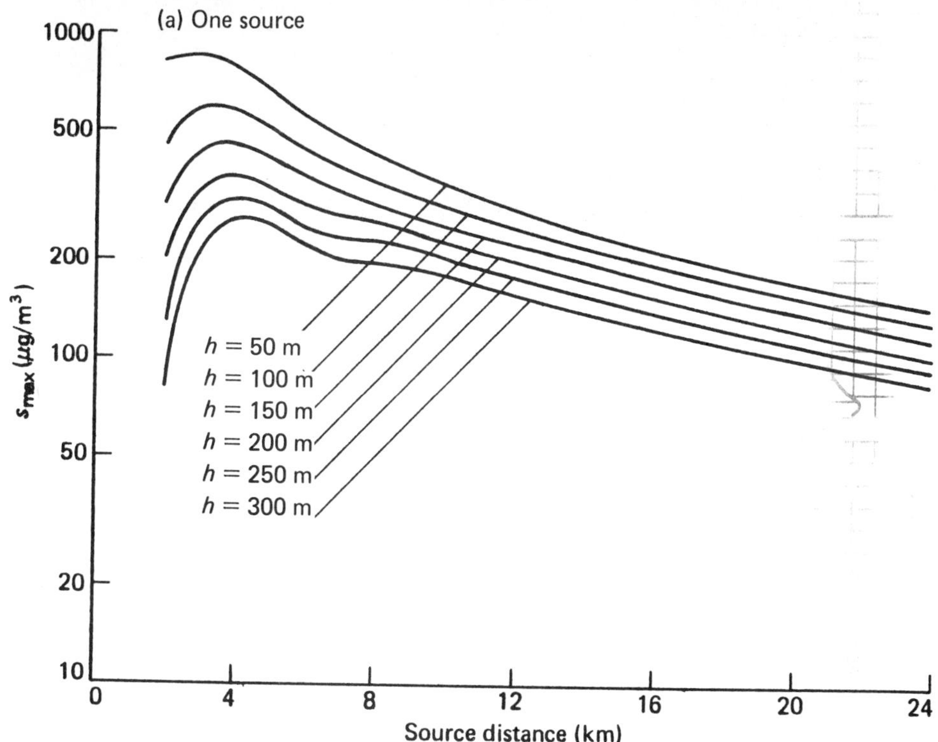

FIGURE 3 *(Continued overleaf; for caption see page 100).*

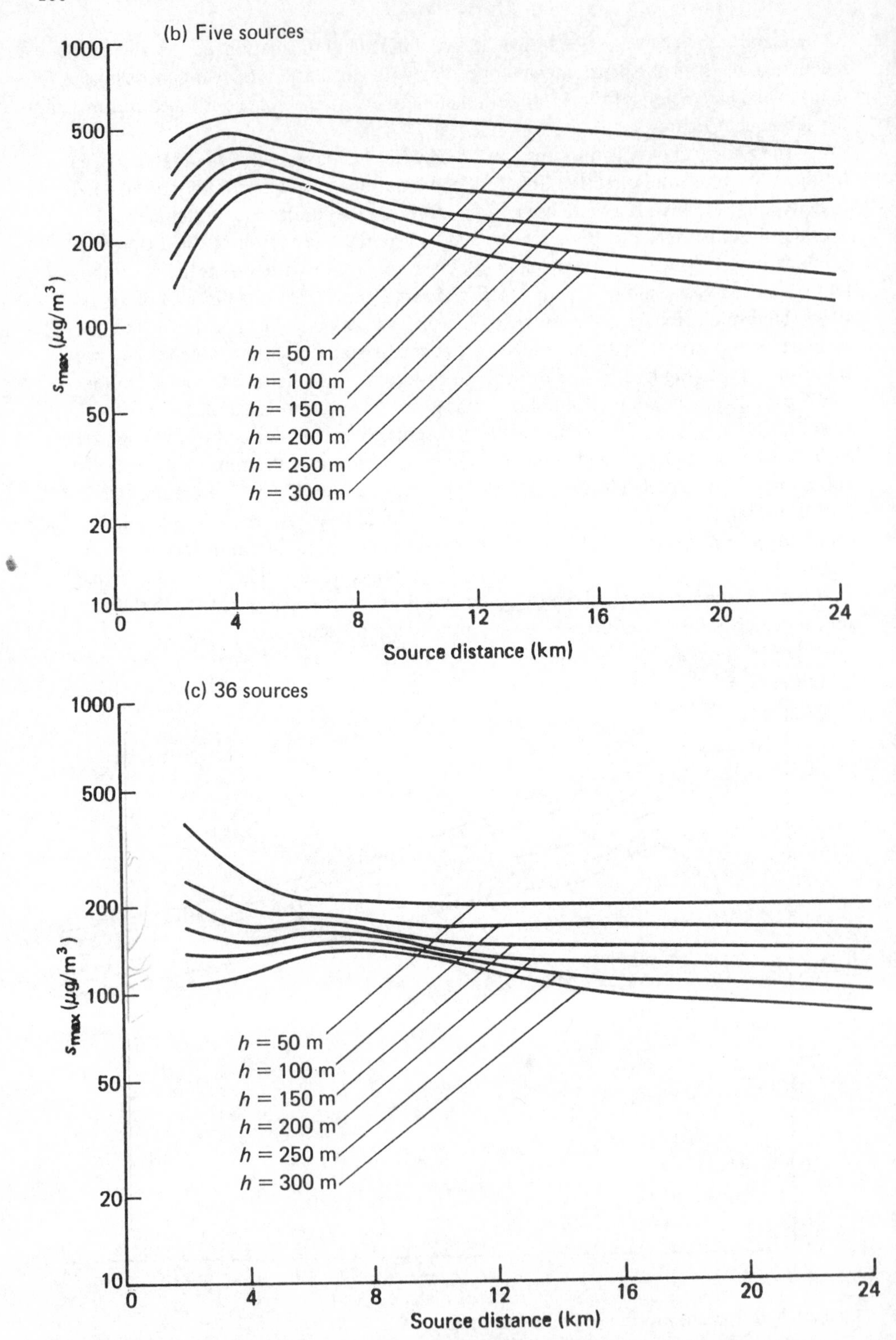

FIGURE 3 Maximum ground-level concentration of SO_2 versus source distance and stack height: (a) one source; (b) five sources; (c) 36 sources.

The parameters of the complete air-pollution climatology are represented as two-dimensional fields in Figures 4 and 5. Firstly the fields of mean values are compared with each other. In the case of only one source the mean values of the concentration for a time period of a normal year are given by Figures 4a and 4b. The difference between Figure 4a (stack height 100 m) and Figure 4b (stack height 300 m) is considerable. The latter case is the more favorable because the island of Neuwerk is fairly well protected; this is also true for the beaches west of Cuxhaven. The 100-m stack height is the most unfavorable because of the very high mean values found very near the island of Neuwerk. The arrangement of five sources seems to be not very different from the single-stack case: stack heights of 100 m (Figure 4c) give almost the same mean values in a slightly changed geometrical form, while stack heights of 300 m (Figure 4d) change the pattern of the mean values only slightly. Comparing these two cases (one stack and five stacks) one should keep in mind that Figures 2 and 3 show that the five-stack arrangement is favorable with regard to maximum ground-level concentrations even though the fields of mean values differ so little from each other. The final decision is made by comparing the 97.5th percentiles (see below). The last field of mean values given in this paper shows the case for an arrangement of 36 sources (each stack 100 m high) (Figure 4e); compared with Figures 4a and 4c, the results are very unfavorable.

In deciding what is the best solution for maximum protection of the environment, the properties of the concentration field statistics already discussed are not sufficient. A comparison of the 97.5th percentiles (Figure 5) gives the final answer. From Figure 5a

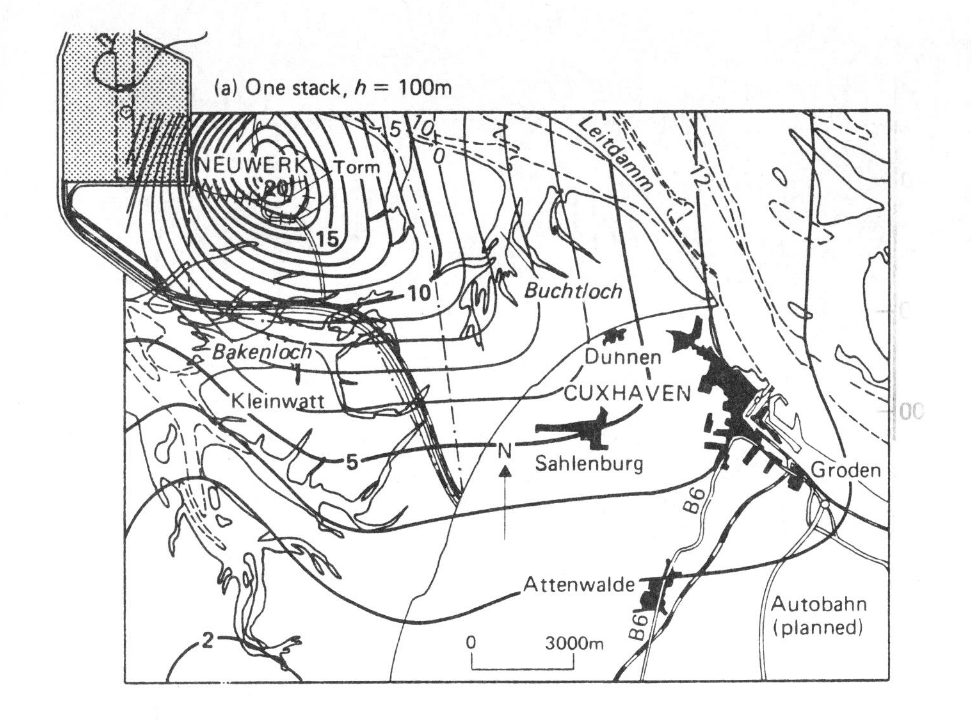

FIGURE 4 *(Continued overleaf; for caption see page 103).*

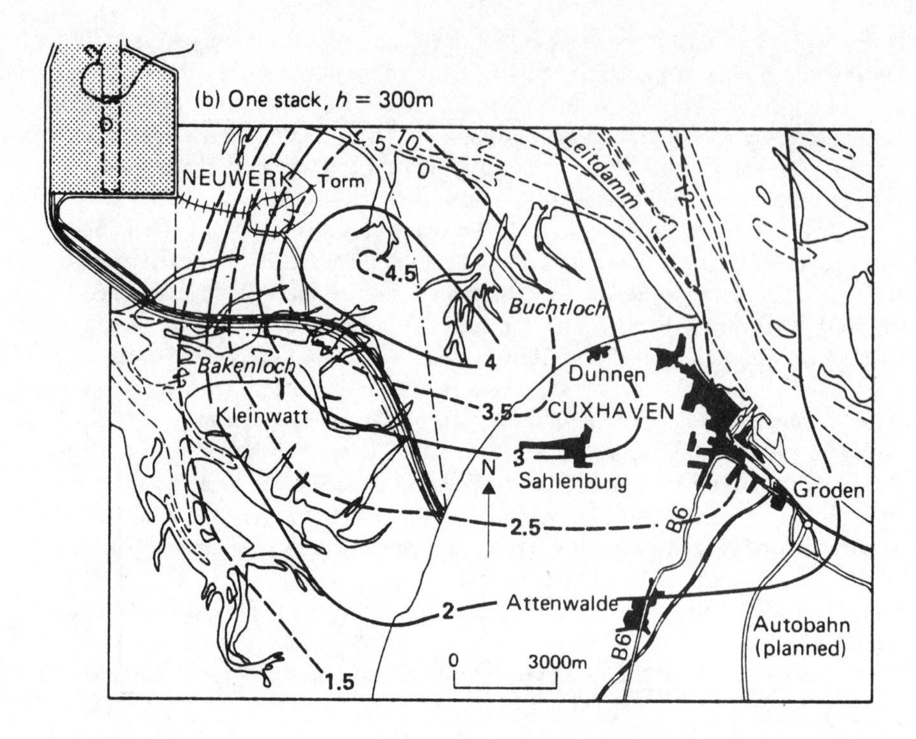

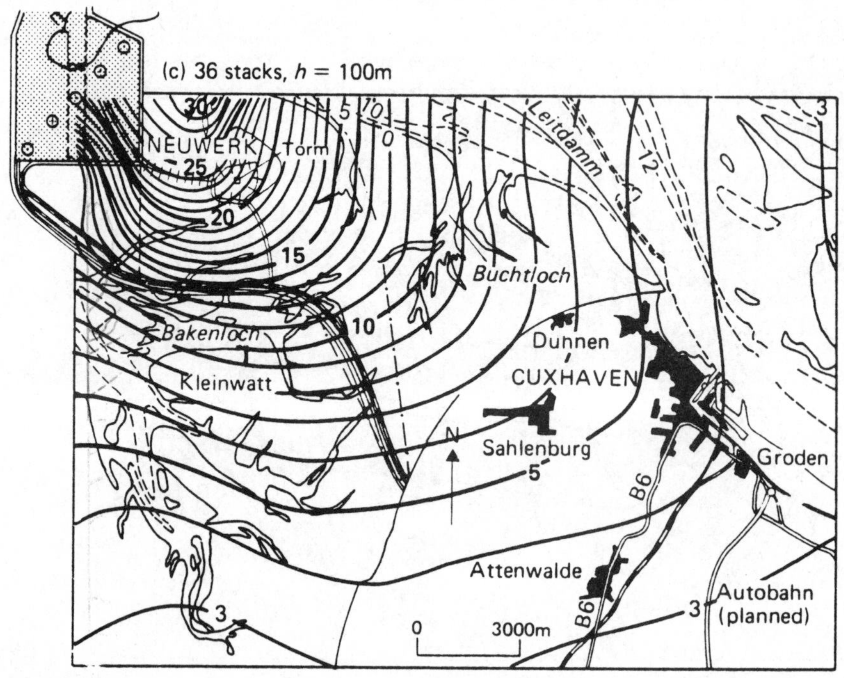

FIGURE 4 *(Continued on facing page; for caption see page 103).*

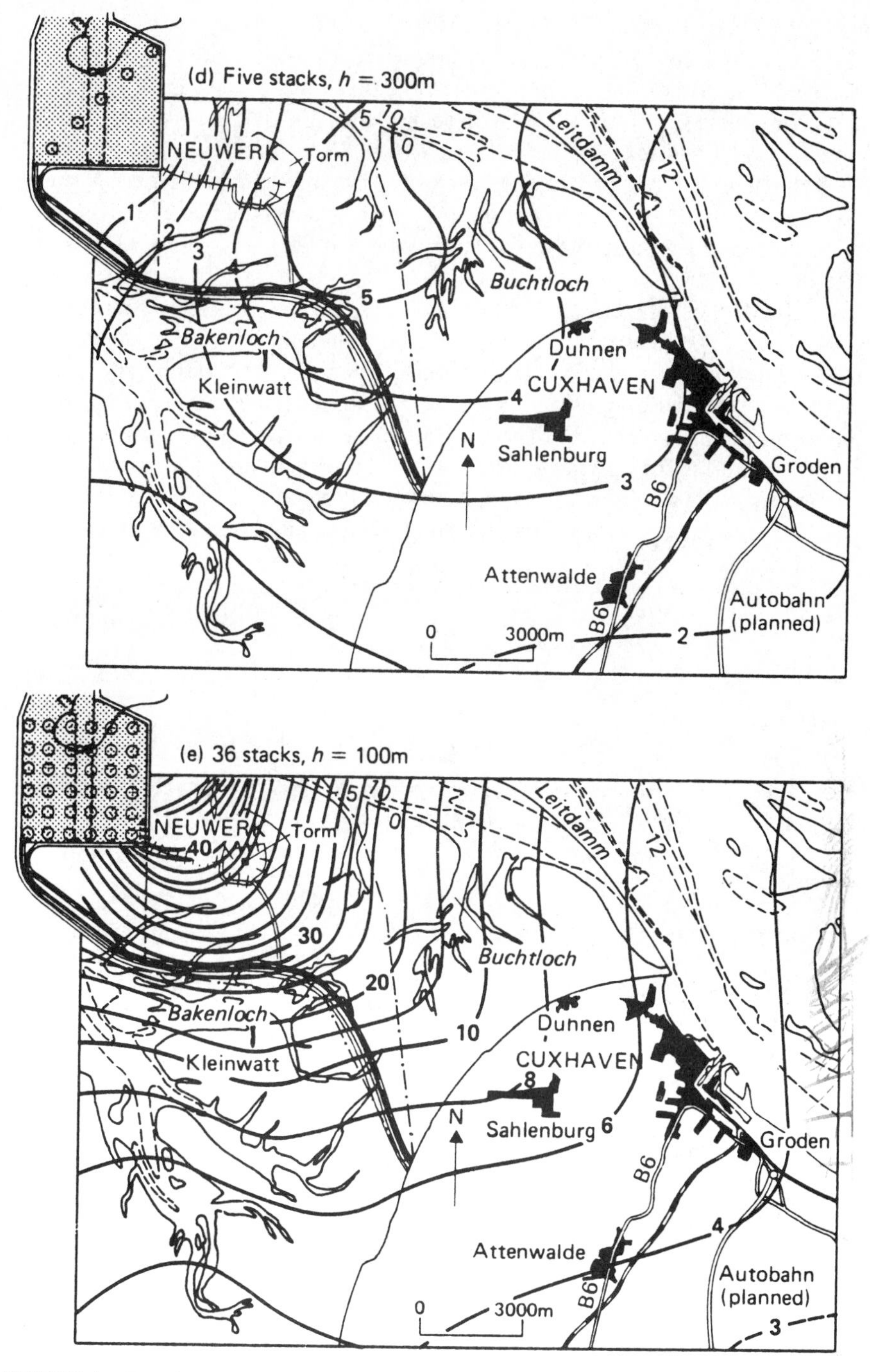

FIGURE 4 Two-dimensional field of the annual mean value of ground-level SO_2 concentration ($\mu g/m^3$): (a) one stack of 100 m; (b) one stack of 300 m; (c) five stacks of 100 m; (d) five stacks of 300 m; (e) 36 stacks of 100 m.

(one source, 100 m high), Figure 5c (five sources, 100 m high), and Figure 5e (36 sources, 100 m high) it can be seen that the five-source arrangement not only gives the lowest 97.5th percentiles at the beaches but also protects the island of Neuwerk sufficiently. The situation becomes even better if Figure 5b (one source, 300 m high) and Figure 5d (five sources, 300 m high) are compared with each other. For one source values between 60 and 70 $\mu g/m^3$ are found at the beaches whereas for five sources values slightly higher than 40 $\mu g/m^3$ are produced. At Neuwerk the difference between the two cases is even greater. For one source a typical 97.5th percentile value is 60 $\mu g/m^3$ compared with a value of 25 $\mu g/m^3$ for five sources.

The selection of experiments shown here indicates that the five-source arrangement with stack heights of 300 m can be considered to be the optimum solution of the stated problem. This solution is not trivial because the many non-linear relationships and the complicated statistical relationships work together in a very complex manner and might have given source arrangements quite different from those we have found in this paper as the optimal solutions. The optimal solution obtained gives very low mean values and 97.5th percentiles in the whole area for a total emission strength of 10 tonne/h. It can be estimated that approximately twice as much could be emitted in the area without violating air-pollution control standards.

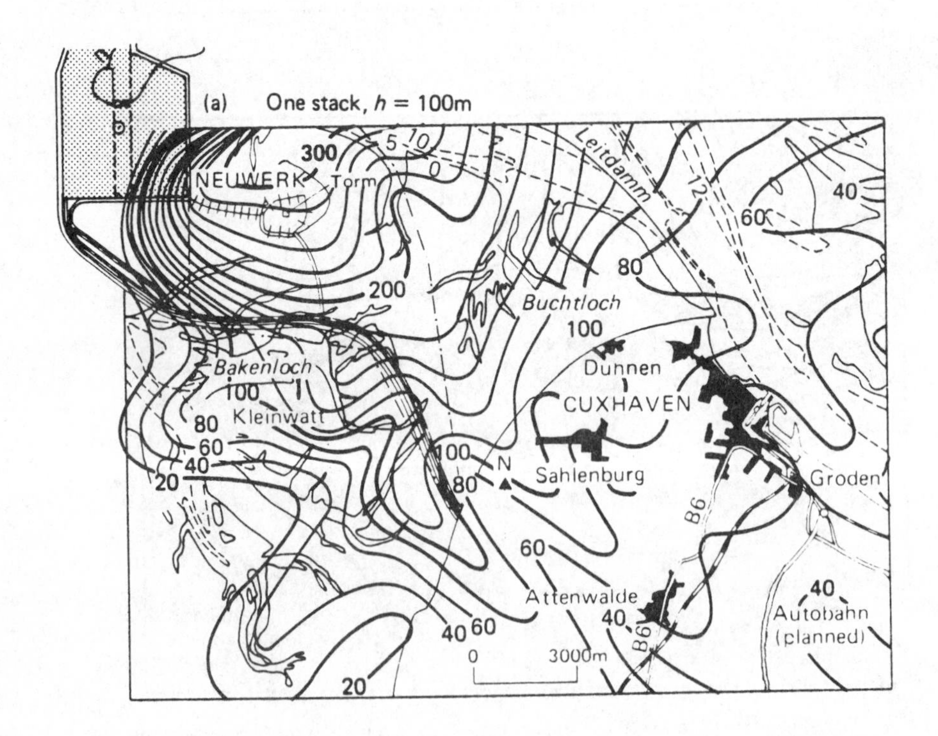

FIGURE 5 *(Continued on facing page; for caption see page 106).*

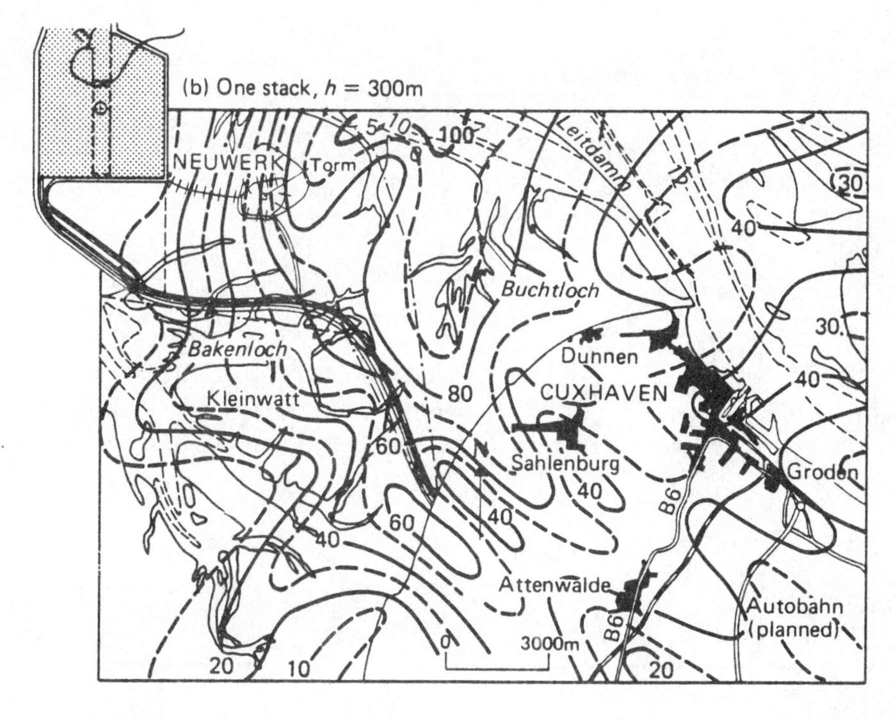

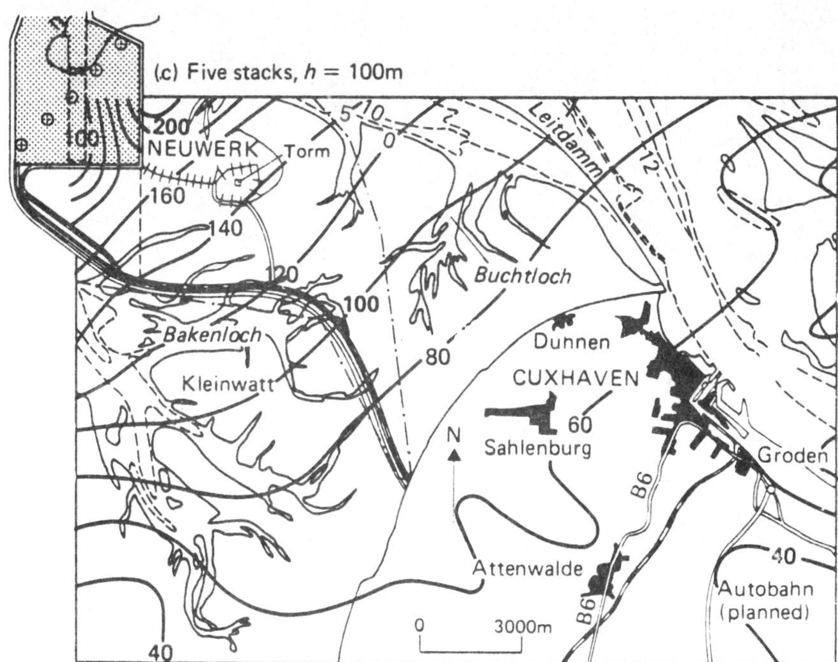

FIGURE 5 *(Continued overleaf; for caption see page 106).*

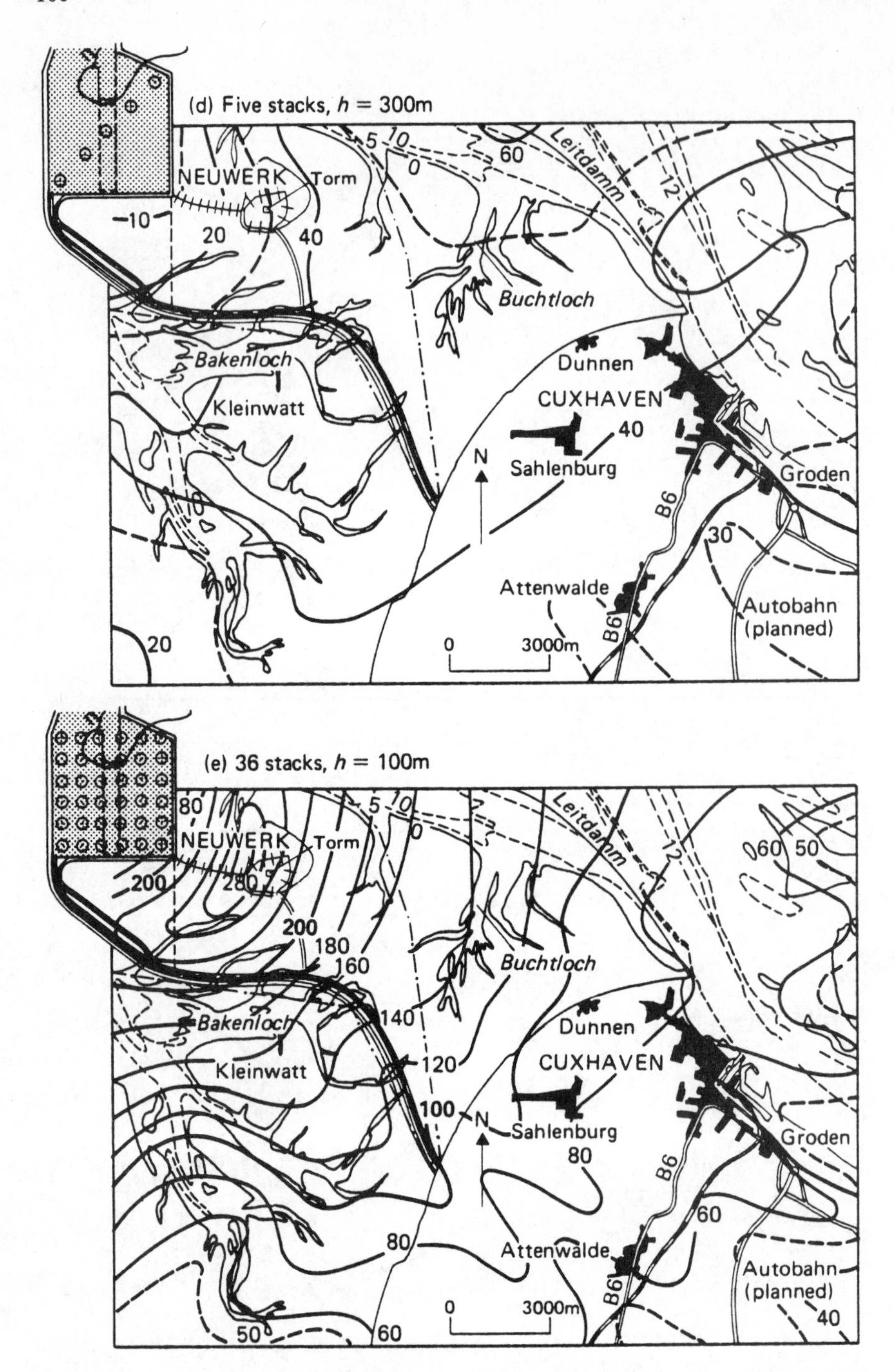

FIGURE 5 97.5th percentile of the annual frequency distribution of ground-level SO_2 concentration ($\mu g/m^3$): (a) one stack of 100 m; (b) one stack of 300 m; (c) five stacks of 100 m; (d) five stacks of 300 m; (e) 36 stacks of 100 m.

ACKNOWLEDGMENTS

The cooperation of my colleagues Professor W. Fett and Dipl. Ing. A. Tamer, who did the programming, and of Dipl. Met. B. Dußler, who provided scientific and technical assistance is gratefully acknowledged.

REFERENCES

Antfang, H. (1969). Die Wind- und Nebelverhältnisse in Elbmündungsgebiet. Hamb. Küstenforsch., 9.

Fett, W. (1974). Simulation der Ausbreitung von Luftverunreinigungen mittels einer Tischrechenanlage. Schr. Reihe Ver., Berlin.

Fortak, H. (1961). Ausbreitung von Staub und Gasen um eine Kontinuierliche Punktuelle in einer bezüglich Windgeschwindigkeit und Austausch geschichteten Atmosphäre. VDI-Forschungsh., H:483.

Fortak, H. (1964). Einbeziehung der Sinkgeschwindigkeit und partiellen Absorption am Erdboden in die Ausbreitungrechnung, speziell in Falle nicht-Fick'scher Diffusion. Rep. Inst. Theor. Meteorol. der FU Berlin, Berlin.

Fortak, H. (1966). Rechnerische Ermittlung der SO_2-Grundbelastung aus Emissionsdaten. Rep. Inst. Theor. Meteorol. der FU Berlin, Berlin.

Fortak, H. (1969). Vergleich von berechneten und gemessenen maximalen Bodenimmissionen und deren Entfernungen von der Quelle für den Fall von Grossemittenten. Staub, 29:493–498.

Fortak, H. (1970). Numerical simulation of the temporal and spatial distribution of urban air pollution concentration. In Proc. Symp. Multiple Source Urban Diffusion Models, RTP, North Carolina.

Fortak, H. (1972). Anwendungsmöglichkeiten von mathematisch–meteorologischen Diffusionmodellen zur Lösung von Fragen der Luftreinhaltung. Herausgegeben von Minister für Arbeit, Gesundheit und Soziales des Landes Nordrhein–Westfalen, Düsseldorf.

Pasquill, F. (1962). Atmospheric Diffusion. Van Nostrand, London.

Singer, I.A. and Smith, M.E. (1966). Atmospheric dispersion at Brookhaven National Laboratory. Int. J. Air Water Pollut., 10:125–135.

Turner, D.B. (1964). A diffusion model for an urban area. J. Appl. Meteorol., 3.

AIR-POLLUTION MODELING OPERATIONS AND THEIR LIMITS

Michel M. Benarie
IRCHA, Vert-le-Petit (France)

1 INTRODUCTION

From the operational point of view the general term "air-pollution modeling" covers three quite distinct types of activity: 1, descriptive; 2, computational; and 3, predictive. For other model systematizations based on input–output or source–receptor relationships, averaging time, or user-oriented considerations, see Benarie (1979).

2 LIMITS OF THE STATISTICAL INFERENCES IN AIR-POLLUTION MODELING

All statistical inference is descriptive. It is a summarization of the data already on record completed by the assumption that the record either is stable or contains trends or cycles which may somehow be extrapolated. Even the most complex statistical model is based on a group of observations; statistical models are therefore essentially empirical. These basic statements encompass all the limitations of these models from the most simple to the mathematically most complex.

Within the class of statistical methods we should clearly distinguish between time-series methods, which do not use meteorological inputs, and regression and similar methods, for which meteorological descriptors are indispensable. In this section we discuss only the time-series methods. Regression methods requiring independent meteorological variables are in fact computational formulas and will be dealt with in the next section.

A pollutant concentration may be considered either as a climatological parameter in the same way as air temperature or humidity or as a complex resultant of "pure" climatological parameters (wind rose, stability wind rose, temperature, etc.). Every population of climatological descriptors is essentially random. Since every function of a random variable is also a random variable the concentrations will be subject to random variation similar to that in a climatological series. We then arrive at the problem of purely statistical weather forecasting. The opinion of meteorologists about this topic is "that past information can never replace present information, nor can present replace past information" (Godske, 1962). This argument is not only scale independent but, as Godske (1962) has shown, it is strengthened by the existence of the widely different time and space scales which occur in macro-, meso-, and micro-meteorology.

The limit of the statistical methods can be perceived most easily by starting with the long-term (climatological) event. The randomness of climatological means also holds true for concentrations. Suppose that we have a method of estimating the average pollutant concentration for next year with the same probability and accuracy with which we can (or cannot) estimate the mean temperature or the total amount of rainfall for next year. There is always an element of chance that next year will be quite exceptional. Only relatively long records show a certain stability of the means; long time series will give useful estimates for the averages of similar long series, but never for the next individual event.

Currently, the Box–Jenkins (BJ) algorithm is considered to be the most sophisticated method for time-series analysis. We therefore submitted the predictive ability of the BJ technique to the following test. After fitting the optimum BJ model to data for 100 days, we compared the extrapolation of the data to the 101st, 102nd, . . . , days with the observations. No fit better than pure chance was obtained. This negative result is important. The time history of the concentrations at any point seems to contain no forward information if the weekly and yearly periodicity are not taken into account, as was the case in our test. Where the BJ technique applied to air pollution does seem to show some predictive ability, a priori meteorological or emission knowledge has been introduced. This is the same conclusion as that reached by Finzi et al. (1977b). Forecasting ability can be obtained only by introducing meteorological inputs such as wind direction and stability. The findings of Finzi et al. (1977a, b; 1979) and Zannetti et al. (1977) concur on this point.

3 COMPUTATIONS AND THEIR LIMITATIONS

Computational models most often arrive by deductive arguments at mathematical formulas which, it is hoped, reflect more or less accurately the physics of the process. To be useful, these formulas need first an adequate amount of meteorological input about the state of the atmosphere (wind velocity and direction, thermal stability, turbulence, etc.) and then similarly detailed data on emissions. The limits of the computational models are therefore threefold: first, in the understanding of the physics; second, in the explicit or implicit simplifying assumptions; and, finally, in the accuracy of the various input parameters. Let us consider the well-known example of the gaussian-plume concept and formula; we can state that, if and when the exact conditions specified by the parameters occur (perhaps the most important cases are when the wind direction is correctly specified, when a plume-rise formula which nearly approximates the real situation is used, or when the stability is evaluated correctly), then the gaussian-plume formulas will give fair approximations of the isopleth contours and the orders of magnitude of the concentrations to be expected. How often and for how long the given set of parameters appears in the surroundings (if they ever do) is beyond the gaussian concept.

By using source-oriented models we attempt to establish a cause-to-effect chain between the emission of a number of sources and the ambient concentration at given locations. The main links of this chain are: 1, knowledge of the source strength; 2, adequate definition of the meteorological parameters; 3, a reliable method for the calculation of the dispersion from inputs 1 and 2; and 4, adequate knowledge of the losses (or formation) of pollutants by chemical or photochemical reactions.

As is well known, a low wind velocity influences mainly points 2 and 3 very strongly. In particular, all known plume-dispersion equations have a singularity near zero wind velocity, and therefore their use at very low velocities becomes suspect.

It should be emphasized that we are not discussing the usefulness of calculations based on plume-dispersion formulas at very low wind speeds. The whole model concept – the causal chain between pollution source and ambient concentration – becomes meaningless when the wind velocity falls below a certain value.

In the terminology of operational research, we are dealing with a multinodal chain. At each node, together with some information, we intoduce more or less random noise. Just such a multinodal chain with noisy input could be used to simulate the outcome of a throw at roulette as follows. Assume that the torque applied to the roulette wheel can be electronically monitored and make the same assumption for the velocity and the angle of the roulette ball. Then apply the known accurate equations of the dynamics of rigid bodies and perform a few more computational steps. You then have the final definitive system to beat Las Vegas!

Obviously you will never be able to do this. By the same logic, multinodal models with the introduction of random noise at every step will not indicate with accuracy the pollutant concentration tomorrow. On the contrary, the more steps (nodes) that are used, the less accurate will be the forecast of the outcome of any individual calculation. Greater sophistication in model building may be a way to improve the precision of averages and of findings about categories, or a way to observe trends, but it seems to be of no use for improving the accuracy of the computation.

This strong statement should not be interpreted as saying that all sophistication is definitively to be rejected. Some very simple one- or two-step schemes show an honorable, if not outstanding, performance. However, if very sophisticated long-chain arguments must be bad, there may still be some intermediate length of operational chain which will give optimum results. Research should be oriented towards methods which are intermediate between the utmost simplicity and noisy sophistication.

The roulette wheel is an example of a mechanical system beyond the reach of mechanical cause-to-effect calculations. However, we shall try to develop this concept gradually by considering a heavy beam supported by an axis of low friction situated near its center of gravity. If the center of gravity is below the axis, the device becomes a sensitive balance. Any perturbation of the balance can be described analytically in terms of oscillation and equilibrium positions. However, if the center of gravity and the rotation axis are made to coincide (they never actually do), the angular position at which the beam will stop can no longer be predicted analytically and the problem becomes one of probability. Somewhere in the process by which the axis approaches the center of gravity the chain of causality breaks down and is replaced by a probability situation. Of course, I do not wish to discuss the fundamentals, since these are well known from the probability calculus; my purpose is only to emphasize that in urban air pollution a situation occurs that is similar to the example of the beam. When the chain of governing equations between cause and effect becomes too long and at each step rather unknown perturbations are introduced, the use of calculus should be abandoned and a new probabilistic approach should be attempted.

This is what occurs in urban air pollution when the wind velocity falls below approximately 3 m/s; above this lower limit atmospheric aerodynamics is a powerful tool, but below or close to it hydrodynamic equations are of as much use as classical mechanics would be for computing on which face a dice will fall. There are two distinct regimes in urban air pollution: one for strong-to-moderate winds and another for light winds in calm conditions.

The difference between urban air-flow conditions with moderate and strong winds and those with light winds, and also the fact that street ventilation changes character when rooftop wind speeds fall to between 2 and 5 m/s, has already been well stressed in the literature.

Insofar as source-oriented models rely on classical analytical equations and a cause-to-effect chain they will behave very poorly in warning systems or in episode-control strategies because generalized and protracted pollution episodes occur mostly during moderate and light winds. In contrast, plume concepts can be quite useful in localizing pollution effects due to point or group sources when the wind speeds are above 3 m/s.

By the same argument source-oriented models, when used as a basis for long-term averages, may be useful if treated with circumspection and provided that light winds and calm periods only occur infrequently. However, if meteorological tables for the urban area of interest indicate that even only 5–10% of the winds have speeds below 2 m/s, then the validity of the concentration distribution as computed by a source-oriented plume model should be questioned. Numerically, these concentrations will be in gross error at the higher levels, which, even if they occur with low frequencies, are the most important levels with regard to effects.

Receptor-oriented models, sometimes with some empirical keying to the source inventory, can be used for warning systems provided that the meteorological parameters are correctly forecast. The vital question is what can be reasonably expected from this kind of forecast?

A major advantage of the finite-difference cell and box models over the plume concept is their flexibility in being able to include wind shear, terrain roughness, and non-linear chemical reactions. Currently, these models are the only means of dealing with rapid atmospheric transformations. Their limitation at present is that even with the most effective of the available computers a compromise must be sought between spatial resolution and computing time. Typical computing cell (grid) sizes range between 1 and 4.5 km. In order to be able to compute in a strictly deterministic way there should be no turbulent eddies smaller than approximately 2 or 9 km. However, in the atmosphere there are eddies as small as 10 cm. Even if we do not go to the extreme and adopt a grid with sides of 100 m, the actual measurement of wind parameters or a numerical solution is out of the question.

The grid size in the kilometer range also introduces an artificial dispersion. A point source inside the cell will be diluted to $10^{-5} - 10^{-6}$ times its original source strength at the grid limit, resulting in quite a large uncertainty in the initial value for succeeding steps.

The following example may also emphasize what can reasonably be expected with regard to the accuracy of air-pollution concentration computation. The average deviation from scheduled arrival times at Paris airport due to weather conditions was only 6 min during 1973 (Benarie, 1976). Flights canceled before departure as well as delays due to technical or commercial reasons are neglected in this statistic. Given that the average flight time was about 3 h this means that the "estimation" was performed with 4% error. Now, these aircraft are driven by thousands of horsepower, are guided by exceptionally skilled crew, and are assisted on the ground by other very competent people and the most powerful computers ever built. If all this complex system results in a 4% relative error, then how can we expect that the calculation of the trajectory of an air parcel driven by its own buoyancy and by turbulent airflow (rather than a jet engine), should perform any better?

4 THE LIMITS SET BY ATMOSPHERIC PREDICTABILITY

The first task, before any attempt at systems analysis, should be to examine whether there is some natural limit to the performance. If such a limit can be clearly defined, many a useless effort (think of the perpetuum mobile) may be avoided. The purpose of the following discussion is the scrutiny of such limits.

We must distinguish quite clearly between forecast and calculation. The latter term is used to denote the operation of taking some formula (e.g. plume, statistical time series, etc.) and then substituting into this formula some assumed (e.g., for the next winter season, etc.) or meteorologically forecast parameters. This is a two-step "if–if–then" process. In contrast, pollution forecasting is episode-centered. An episode means a period with an above-average pollutant concentration. The forecasting of pollution should be defined as a one-step process that estimates a date and a value for a pollutant concentration on the basis of meteorological predictors. The object of the prediction is the pollutant concentration itself.

Air-pollution modeling is based on well-known physical laws related to what is usually called a "volume element", "parcel", or "box" of air. Such an element is a volume of identifiable air that maintains some sort of integrity as it moves around from point to point. The definition of an air parcel depends largely on the scale or size of the process we are considering. A parcel must be large enough to maintain its integrity for a useful period of time and yet small enough to have characteristic properties. Air parcels are labeled by their conservative properties. In meteorology such conservative properties include the water-vapor mixing ratio, the potential temperature, and the absolute vorticity. In air pollution we may add the mixing ratio of inert pollutants. We shall distinguish between the conservation of the identity of air parcels and our ability to simulate or compute their trajectory.

Let us suppose that, at a certain instant, volume elements of air can be marked by tracers, which are "ideal" balloons that are able to follow every motion of the surrounding air and the tracks of which can be observed. Thus each cube is defined by eight balloons (one at each corner of the cube) in the atmosphere. These "mesh particles" will undergo a rapid change in their shapes during the following days; long bands will stretch and finally the development will proceed to a chaotic state where the particles have lost their identity.

All the particles are cubic, i.e. are bounded by square surfaces at $t = 0$. A particle will be considered to have ceased to exist if one of the corner points of the quadrilateral crosses one or both opposite sides in the course of time.

Robinson (1967) found that a particle with a mesh size equal to 300 km should cease to exist within the period 12 h $< t <$ 75 h. Egger (1973), using the data of Kao and Al-Gain (1968) and Kao and Powell (1969) on the large-scale dispersion of clusters of particles in the atmosphere, suggests 45 h $< t <$ 72 h while, using the data of EOLE (Morel, 1972; Larcheveque, 1972) he finds $t \sim 45$ h.

These estimates are upper limits for atmospheric predictability. No numerical forecast model, however designed, can do better than this. Our ability to predict is further limited by the following factors.

One of these arises from the finite representation of the atmospheric fields in the models which makes it impossible to describe scales of motion below the grid scale. Because of the nonlinearity of the hydrodynamic equations, parts of the turbulent energy contained in the subgrid range will appear in another form in the larger scales, thus limiting the predictability of these scales. "It is this last type of uncertainty that is generally felt to be responsible for the limit of predictability of various scales" (Fleming, 1971). Another factor is insufficient knowledge of the initial conditions, e.g. errors in the raw data.

Robinson (1978), considering (weather) forecasting as a hydrodynamic problem and critically examining the application of the Navier-Stokes equation to the atmosphere, arrived at even more categorical conclusions. He showed that the open set of equations which is usually quoted as the result of the averaging process is valid only in the meteorologically trivial circumstance of statistically stationary and homogeneous flow. The only logically valid prediction with this type of equation is one of no change. There is therefore no reason to expect a performance superior to that of statistical or other empirical techniques.

Though it is nowhere clearly stated, a widespread belief prevails in air-pollution circles. This seems to state that for any two time intervals, characterized by an unchanged emission rate and approximately 40 meteorological parameters (wind direction and intensity, thermal gradient, cloudiness, the situation of a given air parcel relative to a front, etc.), if all these parameters are equal, then the pollutant concentrations will also be the same for both time intervals.

By the same logic it could be expected that if 40–60 appropriate parameters were identical then the same form of cumulus cloud would hover over the same quarter of the city. Of course, nobody would dare to assert this as fact. Continuing in this vein, we should not expect that pollution-concentration forecasts will be fully accurate all the time.

Further, it should be remembered in validating a model that correlation does not imply causation.

Lack of reasonable model correspondence with the historical picture speaks strongly for invalidation. However the achievement of such correspondence, while gratifying, really only lets us move on to the next step in the process. It does not "validate" anything, and it tells us precious little about how much we should believe in the model as a predictor of the future. This is true because practically any complex model can be "tuned" to fit practically any given pattern of historical data. Since the causal structure of such a tuned model need have nothing in common with that of the real world, its predictions under new conditions are highly unlikely to correspond to reality. This situation is similar to the well-recognized danger of extrapolating (or for that matter interpolating) from general polynomial regressions to situations outside the range of observations (Holling, 1978).

REFERENCES

Benarie, M. (1976). Urban air pollution modeling without computers. US Environmental Protection Agency Publ. EPS-600/4-76-055. NTIS-PB-262393. National Technical Information Service, Springfield, Virginia, USA.

Benarie, M. (1979). Urban Air Pollution Modeling. Macmillan, London.

Egger, J. (1973). On the determination of an upper limit of atmospheric predictability. Tellus, 25: 435–443.

Finzi, G., Fronza, G., Rinaldi, S., and Zannetti, P. (1977a). Modeling and forecast of the dosage population product in Venice. In Proc. IFAC Symp. on Environmental Systems Planning, Design, and Control, Kyoto, Japan.

Finzi, G., Fronza, G., and Spirito, A. (1977b). Univariate stochastic models and real-time predictors of daily SO_2 pollution in Milan. In Proc. NATO–CCMS Int. Tech. Meet. on Air Pollution Modelling and its Application, 8th, Louvain-la-Neuve, Belgium.

Finzi, G., Zannetti, P., Fronza, G., and Rinaldi, S. (1979). Real-time prediction of SO_2 concentration in the Venetian Lagoon area. Atmos. Environ., 13: 1249–1255.

Fleming, R.J. (1971). On stochastic dynamic prediction. II: Predictability and utility. Mon. Weather Rev., 99: 927–938.

Godske, C.L. (1962). Methods of statistics and some applications to climatology. In Statistical Analysis and Prognosis in Meteorology. World Meteorological Organization, Geneva. Tech. Note 71, 9–86.

Holling, C.S. (1978). Adaptive Environmental Assessment and Management. Wiley, New York.

Kao, S.K. and Al-Gain, A. (1968). Large-scale dispersion of clusters of particles in the atmosphere. J. Atmos. Sci. 25: 214–221.

Kao, S.K. and Powell, D. (1969). Large-scale dispersion of clusters of particles in the atmosphere. II: Stratosphere. J. Atmos. Sci. 26: 734–740.

Larcheveque, P. (1972). Turbulent disperson – EOLE experiment. Committee on Space Research XV, Madrid.

Morel, P. (1972). Satellite techniques for automatic platform location and data relay. Committee on Space Research XV, Madrid.

Robinson, G.D. (1967). Some current projects for global meteorological observation and experiment. Q. J. Res. Meteorol. Soc., 93: 409–418.

Robinson, G.D. (1978). Weather and climate forecasting as problems in hydrodynamics. Mon. Weather Rev., 106: 448–457.

Zannetti, P., Finzi, G., Fronza, G., and Rinaldi, S. (1977). Time-series analysis of Venice air quality data. In Proc. IFAC Symp. on Environmental Systems Planning, Design, and Control, Kyoto, Japan.

AIR-QUALITY MODELS, PARTICULARLY FOR THE MESOSCALE

L.E. Olsson
Swedish Meteorological and Hydrological Institute, Stockholm (Sweden)

1 INTRODUCTION

Early air-quality modeling focused on the local scale. During the last decade models of long-range air-pollution transport have been developed, but it is only in the last few years that air-quality modeling for what in this project is defined as the mesoscale (10–300 km) has been developed.

In addition to a need for further development of physical and numerical models there is also a need for studies of how mesoscale circulations such as land- and sea-breeze systems influence air-pollution dispersion. As we move from the local scale to the mesoscale we have to pay increased attention to various physical and chemical transformations. This intermediate scale is of great importance for several reasons: increased emissions from single point sources, higher stacks, increased attention to problems related to deposition, air-pollution transport across international boundaries, etc.

The idea of organizing a joint Nordic research project for the development of "unified methods for the calculation of dispersion of air pollution in the mesoscale" was first discussed at the Nordic Symposium on Urban Air-Pollution Modeling in Denmark in 1973. In 1975 the Nordic Research Council initiated a joint research project between 11 institutions in Denmark, Finland, Norway, and Sweden under the project title "Mesoscale Dispersion Modeling". At the same time the Nordic Research Council initiated a project on the development of a "Nordic Ventilation Climatology", which was concluded with a final report in the fall of 1977. The two projects were operated in close coordination, with several joint meetings between the scientists involved.

In 1979 the final report on the Mesoscale Dispersion Project was published and in the present paper we shall attempt to summarize the findings of the project. Research and development moved fast during the project period (1975–1978) and the problems of mesoscale dispersion proved to be very complex. Here we limit ourselves to a general overview of the project and to some highlights in the final results. Several of the subprojects have been published in scientific journals and have been collected in three annual reports. Some of the material is available in English.

Instead of reaching the original goal of the project, i.e. establishing "a unified modeling technique to be applied in the Nordic countries", the project defined the various problem areas related to mesoscale modeling. The final report can be considered as a presentation of the state of the art of mesoscale modeling as of 1978.

2 A USER-ORIENTED PRESENTATION

One of the most severe problems connected with air-quality modeling is the difficulty of communication between modelers and users. The first part of the report is addressed to various users, i.e. people who do not have professional knowledge of air-pollution meteorology or numerical modeling but who have to rely on the results obtained from air-quality models. The user might be a land-use planner, an engineer in charge of plant operation, an environmental consultant, or any decision maker.

In order to facilitate communication an attempt is made to define and clarify the following:

- the concept of modeling;
- the typical applications of dispersion modeling (Table 1);
- factors to be considered in selecting an appropriate model for a specific problem;
- the accuracy of model results and its dependence on input data, model sophistication, time and space resolution, etc.;
- model verification;
- the need for computing facilities;
- the various ways of presenting model results, e.g. averages, extremes, high-concentration maps, time and space dependences, frequency distributions, various types of forecasts of air quality;
- the classification of various dispersion models (Table 2);
- various modeling types, e.g. a point source, multiple point sources, area sources, etc.

TABLE 1 Typical applications for air-quality models.

1. Point sources	8. Photochemical oxidants
2. Large point sources	9. Distant sources
3. Stack height	10. Land-use planning
4. Emission control	11. Traffic planning
5. Accidental release	12. Planning of measurement programs
6. Deposition	13. Analyses of measurements/trends
7. Odor	14. Episode forecasting

Primarily for the benefit of the various users of air-quality models, but also as information for model developers and scientists, we have attempted in Table 3 to summarize schematically air-quality modeling and its application in the Nordic countries with cross-references to model applications, model types, and appropriate papers and reports. Although it is impossible to use the information in Tables 1–3 without detailed information and a full list of references the general idea should be clear.

TABLE 2 Classification of dispersion models according to the criteria: *A*, complexity of meteorology; *B*, type of source of major concern; *C*, desired averaging time; *D*, type and characteristics of pollutant.

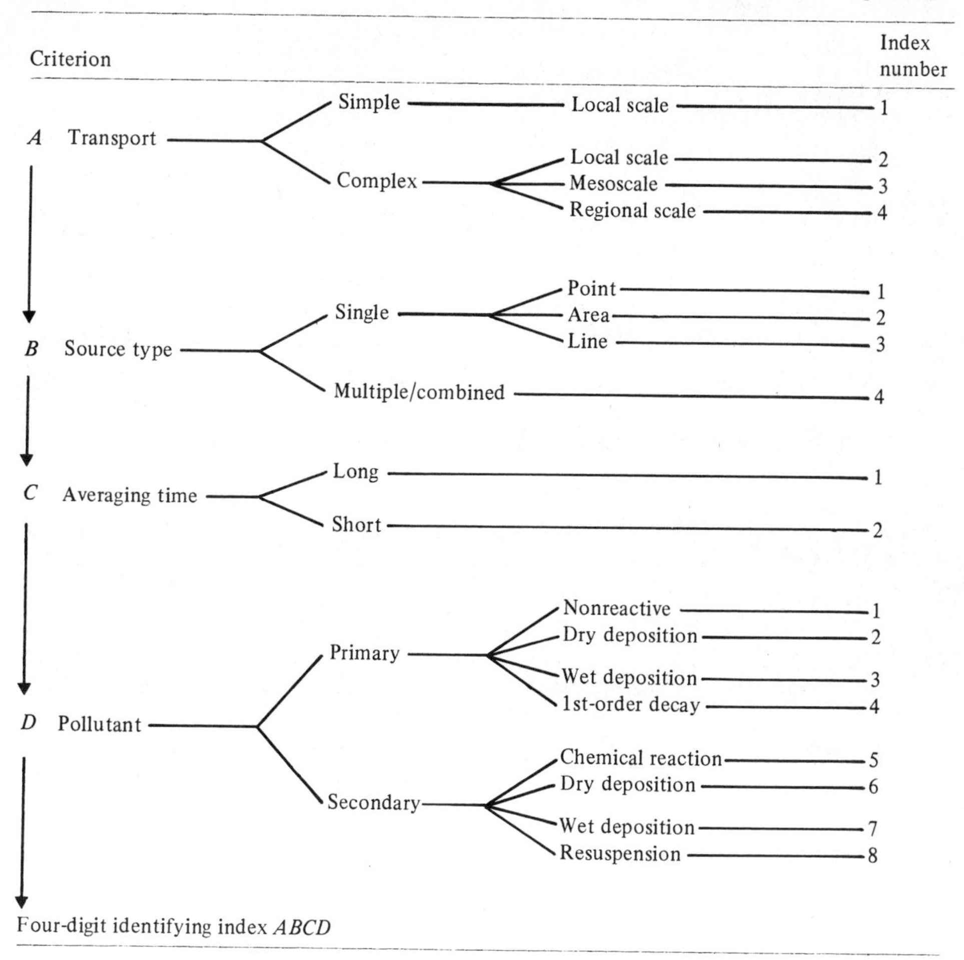

Tables 1–3 are designed to be used as shown in the following example.

If you are charged with estimating the environmental impact of *nonreactive* air pollution from a *large point source* on the *local scale* you have *application type 2* (Table 1) and from the model "classification tree" (Table 2) an *index 11C1*. In the reference table (Table 3) you find a suggestion that you can use gaussian models, and there are several references to papers describing such models; such papers and reports have been presented in Finland, Norway, and Sweden. However, if you want to make estimates for *short averaging time* in *mesoscale* you arrive at *index 3121* and you need to consider trajectory–puff models. You will find some relevant discussion of such models in references 26 and 120, but there do not seem to be any references to applications of these modeling techniques in the Nordic countries.

TABLE 3 The models available, with details both of their general application and of their application to the Nordic countries in particular.

Model type	References to model description[a]	Possible application: Application[b]	Possible application: Classification index[c]	Examples of application: Application[b]	Examples of application: Classification index[c]	References to available reports: Denmark	Finland	Norway	Sweden
Single source									
Gaussian stationary	63, 127, 115, 79, 22, 64, 14	1–7 10 12–14	11*C*1 11*C*2 11*C*3	1,2	11*C*1–4		31	108	88
				3	1121	22	32	38	33
				4	11*C*1			109	
				5	11*C*4	100	103		72
				6	11*C*2				91, 89
				7	1121				1, 14, 45
Numerical integration of diffusion equation	4, 39	1, 2, 5–14	*ABCD*	6	1322	4, 96, 5		39	
				11	1421				
Trajectory model without dispersion	56	1, 2, 5, 9	*A*111	5	41*C*4				90
				9	4*B*24	92			87
Trajectory model with dispersion (puff–trajectory)	26, 120, Appendix B	1, 2, 5–14	*A*1*CD*	9			61		
Multiple source									
Box model	10, 14, 28, 37, 52, 97	10, 12, 13	14*C*1	10	1411			51	
Dispersion eqn. (gaussian model)	15, 48, 49, 71, 93, 104	1–7 10, 12–14	14*C*1	10	14*C*1	93	55	51, 110,	98, 83,
				11	13*C*1	125		74	99, 124
Numerical integration of diffusion eqn.	5, 19, 20, 46, 48, 86, 94, 95, 96 (Appendix C)	1, 5, 6 9–14	*ABCD*	1	3221			110	
				8	3225			50	
				9	42*C*1–4	86, 94	78		
				10	32*C*1			47	
Trajectory model	53, 78	1, 2, 5–7, 9, 10	*A*4*C*1–4	9	4411			29	
Statistical regression model	13	10, 13, 14	1411	13	1411				13

[a]The references given correspond to references in the original report on the Mesoscale Dispersion Project.
[b]The numbers correspond to the applications listed in Table 1.
[c]The numbers correspond to the numbers at the far right-hand side of Table 2.

3 MESOSCALE MODELING

The second part of the report discussed earlier focuses on the modeling of mesoscale dispersion of air pollutants. A summary of the approach is presented in Figure 1. The first column indicates what happens in the "real atmosphere". The second column illustrates what type of measured or observed input data are needed for air-quality models. Despite the vast amount of environmental data, e.g. at the national meteorological services, one is frequently faced with a severe shortage of adequate and properly representative input data. Meteorological and climatological measurements and observations have traditionally been made for purposes other than environmental protection and air-quality modeling. Furthermore, since we are dealing with the mesoscale the time and space resolutions in the required input data demand a radically new and more system-oriented approach in measurements and observations.

To observe and measure mesoscale atmospheric circulations with conventional techniques is in general very expensive. Remote-sensing techniques for measurements of winds and stability variation in the planetary boundary layer will open up new avenues in this area. Efficient data-acquisition techniques based on automatic weather stations with microprocessors will also help in the collection of more-adequate mesoscale information. In Sweden we now have five 300-m telecommunications masts and several lower masts equipped with automatic systems for measurements of wind and temperature profiles. These measurements are primarily made for wind-energy purposes but will in fact be of great help in improving our inputs to dispersion models.

The rapid development in the field of boundary-layer modeling has also improved the possibilities for describing the wind and diffusion features on the mesoscale. This is part of what is indicated in column 3 of Figure 1, which might be likened to the "objective analyses" used in connection with numerical weather-prediction systems. In this case boundary-layer models might be used in a "passive mode" in relation to the air-quality model, e.g. to generate wind fields and fields of the diffusion coefficients (K) and to calculate mixing heights. A boundary-layer model can of course be more or less sophisticated; i.e. it can include the physical description of the atmosphere in more or less detail and also be more or less tied to measurements and observations.

Statistically based analysis methods and various interpolation methods, e.g. Gandin's "optimal interpolation", are frequently used to create, for example, wind fields. Indirect methods based on wind, temperature, and insolation, are used to generate diffusion parameters.

An important parameter even in mesoscale air-quality modeling is the "plume rise". Several models have been developed and most have the form $\Delta h = A/u_s$, where A is a function of the "exhaust speed" and "excess heat" in the plume and u_s is the wind speed at the top of the stack. The most frequently used models are summarized in Table 4.

In all dispersion calculations it seems reasonable to define dry deposition as the flux of pollutants from the air to the surface. The most straightforward approach is to use the concept of "deposition speed". However, in impact studies it is frequently more convenient to use the inverse of deposition speed, i.e. the total deposition resistance. The total resistance (r) can be written as the sum of various resistances:

$$r = r_a + r_b + r_s$$

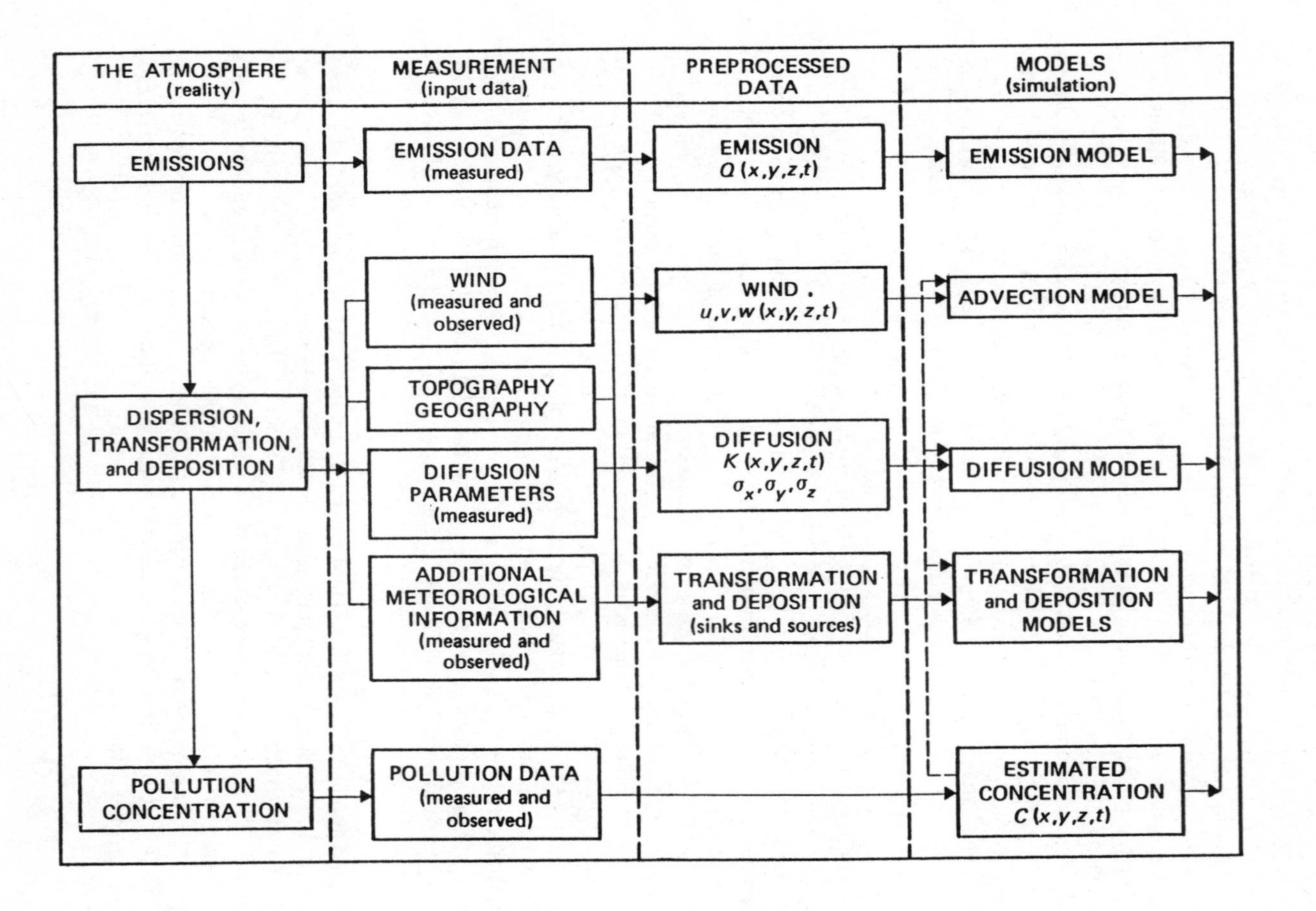

FIGURE 1 Flow scheme suggesting modules or elements to be considered in assembling an air-pollution dispersion model.

TABLE 4 Plume-rise formulas versus pollution situations.

Source	Plume-rise equation	Reference[a]
Small emission, single pipe, domestic	Holland	59
Industrial, warm emission (< 30 MW)	Stümke	121
Small and intermediate heating plants, industrial	Bringfelt	12
Power plants, hot emission (> 30 MW)	Briggs	9
Radioactive emission	Gifford	36

[a]The reference numbers correspond to references in the original report on the Mesoscale Dispersion Project.

where r_a is the aerodynamic resistance, r_b is the boundary resistance, and r_s is the surface resistance. Depending on what type of dispersion model is used one might use a "source-depletion" or "surface-depletion" formulation in the deposition model. From a physical point of view surface-depletion models are more satisfactory.

Several models have been presented for wet deposition. The efficiency of wet deposition depends on parameters such as the precipitation intensity, the drop size, the pH in the initial drop, the solubility of gas in the drop, and the diffusion from the drop. Wet deposition is a difficult phenomenon to include in any model. Several experimental studies of wet sulfur deposition have been presented in recent years for both the local scale and the regional scale. Attempts have been made to explain the great differences in characteristic residence distance for sulfur released during periods of precipitation, and there seem to be systematic differences depending on the air-mass history.

As we move from the local scale to the mesoscale, chemical reactions in the atmosphere become more important. The degree of chemical transformation of pollutants in the atmosphere is dependent on insolation, temperature, etc., and varies from the gas to the liquid phase. Great attention has been paid during the last few years to the photochemical formation of ozone and other oxidants in the lower atmosphere. The transformation of SO_2 to sulfate occurs both in the gas and the liquid phase and the transformation rate is strongly dependent on factors such as relative humidity, temperature, and pollution concentration.

One of the conclusions of the reported project is that in order to improve air-quality modeling on the mesoscale much work will have to be done in the field of atmospheric chemistry.

The most feasible way to establish a mesoscale dispersion model is to develop relatively independent submodels, each simulating processes of importance. Even though the work involved in assembling a complete mesoscale dispersion model requires relatively great efforts the project has shown that such a model can be assembled. Subroutines and available submodels have been suggested and some examples of how complete models can be formulated are offered in appendixes to the report.

The resulting estimated concentrations can only be as good as the input data used and the physics and mathematics used in the various submodels. It is felt that although further improvements may still be required, air-quality modeling for the mesoscale can already be of great value as a tool in environmental planning and protection.

GENERALIZING THE CONCEPT AND FACTORS OF AIR-QUALITY MANAGEMENT

D.J. Szepesi
Hungarian Meteorological Service, Budapest (Hungary)

1 INTRODUCTION

Ideally, the fate of atmospheric trace constituents would be studied by using a global-scale transmission model comprising smaller-scale submodels and taking into account the input of all natural and man-made sources, and the three-dimensional and time-dependent transport, dispersal, removal, and transformation mechanisms of the trace constituents correctly and in sufficient detail. It is well known that the state of the art is far behind the present requirements in all aspects and, although modeling techniques have reached a certain level of success, their general applicability is not yet satisfactory.

To investigate smaller-scale, mostly anthropogenic air-pollution processes, researchers and air-quality managers can either neglect the pollution effects resulting from larger-scale processes or take them into account as background pollution. This means that the need to define and use the background-pollution concept and values will exist until satisfactory global-scale modeling has been developed. This could be the case for many years to come. For simpler studies, when complicated modeling is not feasible, practical solutions can and will be achieved only by making drastic simplifications similar to those outlined above.

2 AIR-POLLUTION SCALES

The concept of background pollution, which originated in the 1960s, is developing from one of climatological–administrative character to one of synoptic–operational character. The basic goal of the climatological–administrative background concept was to collect reliable data on the effect of man-made pollution on the atmosphere and to make global climate studies. Later the program was extended to provide data for studies of air–surface interchange and transport within and between regimes, but these developments did not fit within the original framework.

Synoptic–operational concepts were developed at the national and continental levels which, not being coordinated internationally, were characterized by many contradictory elements as far as goals, terminology, and scales were concerned (see Table 1).

A possible way out of the present situation is to transform the earlier climatological concept into a synoptic–operational one which, at the same time, retains all the positive elements of the earlier system and ensures the continuation of the ongoing WMO program to achieve a long series of air- and precipitation-chemistry data. On the other hand, only a synoptic–operational system and the modeling based on it, is capable of making a distinction between natural, short-lived, and long-lived atmospheric trace constituents, a basic requirement of the whole exercise.

Firstly, it is necessary to define air-pollution processes (or regimes). Considering their transport, dispersion, removal, transformation, emission, and air-quality components, air-pollution processes are integral and persistent atmospheric mechanisms which evolve through characteristic stages in time and space.

3 SCHEME OF BACKGROUND POLLUTION

As air-pollution control in cities and industrial areas becomes more and more effective, the interest in air-pollution studies is shifting in many countries from local and urban scales to regional, continental, and global problems. However, though the use of passive-control strategies, such as the increase of stack height, is often an effective tool for decreasing local-scale pollution, it can actually increase the effect of the larger-scale pollution processes. The different aspects of pollution processes will now be considered.

Before establishing new scales for air-pollution processes, it is necessary to review the existing meteorological and climatological scales, shown in Figure 1. A closer inspection of Figure 1 reveals that considerable disagreement exists regarding the identified phenomena and their scales. Besides, air-pollution processes must be identified by both their pollution and meteorological aspects. On the basis of these considerations it can be concluded that air-pollution processes should not be classified according to purely meteorological scales alone.

A scheme of the proposed new system of scales of air-pollution processes is shown in Figure 2. The basic idea of the new system is as follows:

(a) The system is receptor oriented.

(b) Air-pollution processes are defined as functions of distances upwind of the receptor.

(c) Background pollution from a larger-scale pollution process is superimposed on the polluting effect of the smaller-scale process, for example, the continental background plus the regional polluting effect give the regional background pollution.

(d) Receptor points for the measurement of concentration levels caused by a pollution process of a given scale should be located so as to ensure that they are not affected by other (smaller-scale) processes; for example, continental-level pollutant concentration can be measured at a certain point if pollution processes of regional, urban, and local scale are absent for a distance of 100–200 km around the receptor.

(e) The ratios of the natural plus long-lived trace constituents to the short-lived ones change considerably with increasing scale.

(f) Monthly mean sulfur-dioxide and particulate-sulfate concentrations show a variation of roughly two orders of magnitude on going from the local to the global level.

Scale suggested by	Scale (km): 0.1 – 1 – 10 – 100 – 1000 – 10 000
Hall and Holloway (1955)	0.5–10 micro; 10–300 meso; 150–1500 macro
Flohn (1959)	0.001–0.1 micro; 1–10 local; 100–500 regional; 1000–5000 macro; > 10 000 global
Takahashi (1969)	< 10 micro; 1–50 meso; 100–10 000 synoptic
Yoshino (1961)	< 0.1 micro; 0.01–10 local; 1–200 meso; 200–20 000 macro
Mason (1970)	0.1–1 micro; 1–20 convective; 20–500 meso; 500–5000 synoptic; > 5000 planetary
Szepesi (1982, this paper)	0–10 local; 0–30 urban; 30–200 regional; 200–3000 continental; > 3000 global

FIGURE 1 Existing meteorological and climatological scales (km) (source: Yoshino, 1975).

TABLE 1 History of air-pollution scales (Air Quality Control Region).

Scale	Levels					
Climatological scales						
WMO (1969)	Local		Regional			Global
Objectives	Protect human health		Anthropogenic effect on atmosphere			Climate studies
Overall scales	←*Impact level (WHO)*→		← *Background level (WMO)* →			
WMO (1976, 1978)			Regional	Regional with extended background		Global
Objectives			Determine (a) Long-term changes in atmospheric composition due to changes in regional land-use practices and man-induced activities (b) Air–surface exchange and atmospheric transport within regimes characterized by significant man-made influences	Determine (a) Transport and deposition of potentially toxic substances (b) Air–surface within and atmospheric transport between large-scale areas characterized by different biomes and different sea surface conditions		Determine (a) Global inventories and their trends for climate studies (b) Latitudinal transport for global biogeophysical modeling
Operational scales						
Rossano and Thielke (1976)	←Local→		Regional	Statewide	Continental	Global
Hidy et al. (1977)	Local Microscale 0–10 km	Urban Mesoscale 10–100 km	Regional Synoptic 1000 km			Global Planetary > 1000 km
L. Smith, personal communication (1977)	Short range < 50 km	Subregional AQCR 50–200 km	Regional State(s) 200–500 km	Long range States > 500 km		
Proposed new scales						
	Local	Urban	Regional	Continental		Global
	← Short range →		← Long range →			
Distance upwind from receptor (km)	0–10 (20)	0–30 (100)	30–200 (300)	200–3000		> 3000
Sampling height (m)	2–600	2–1000	2–1500	2–3000		2–10,000

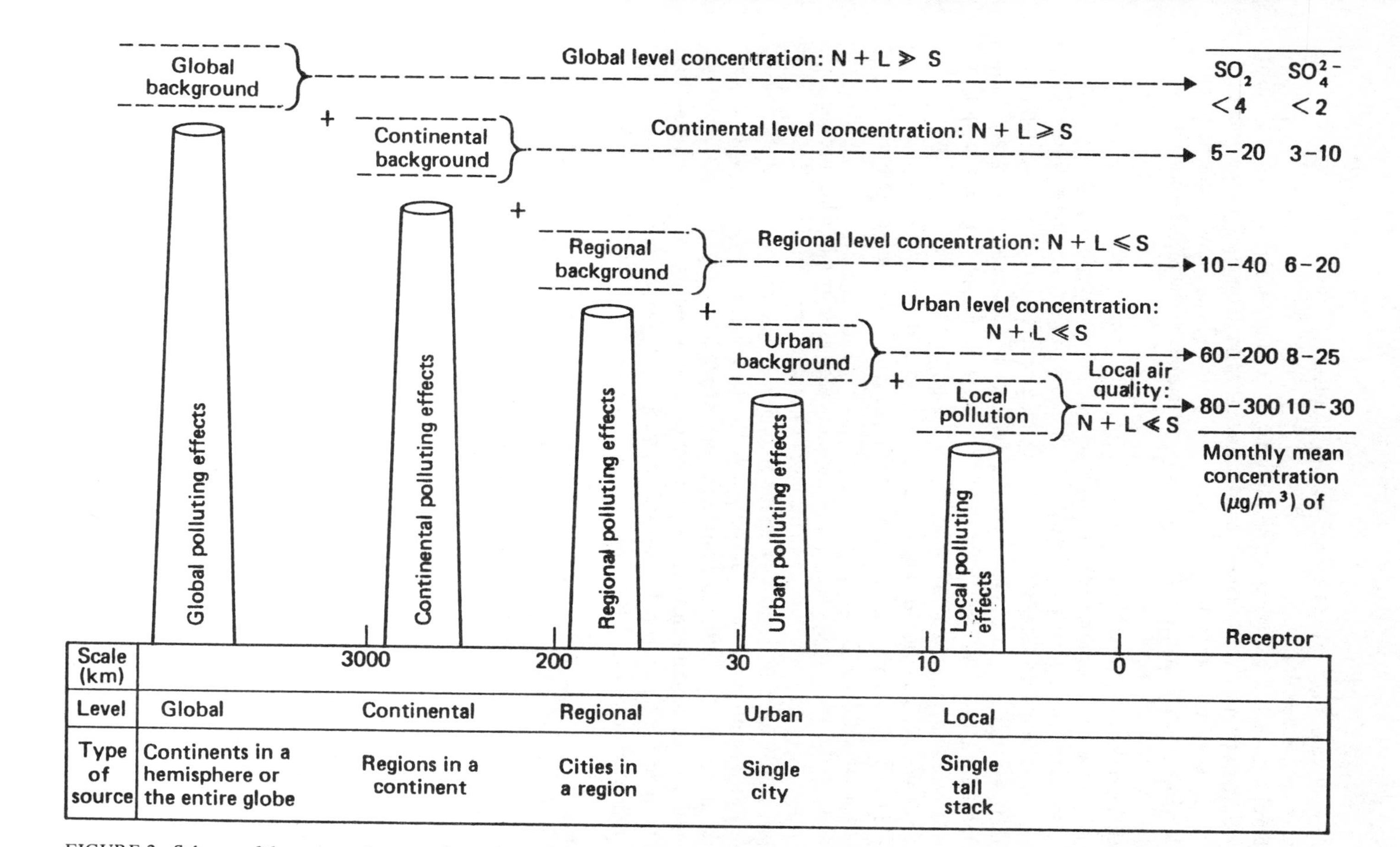

FIGURE 2 Scheme of the proposed new scaling system for air-pollution processes [average values reported by Georgii (1978), Hidy et al. (1978), Mészáros (1977, 1978), Ottar (1978), and Perhac (1978)]. Characteristic relative concentrations of natural (N), long-lived (L), and short-lived (S) trace constituents are shown for each level.

4 SCALES OF AIR-POLLUTION PROCESSES

Air-pollution processes will now be described in terms of their source, scale, and background characteristics.

Local-scale pollution originates from individual point, line, or area sources, the maximum polluting effect of which can be easily distinguished from the background pollution of a larger-scale process within 10 or 20 km downwind of the source. Local-scale pollution is considered allowable if it meets short- and long-term air-quality standards based on health and economic considerations.

Urban-scale pollution originates from multiple area, line, and point sources. The distinction of urban-scale pollution from local- or regional-scale processes is justified by the fact that this is the only scale where a high density of relatively homogeneous anthropogenic emission can be found over a very large area. The horizontal scale of urban pollution processes is 30–100 km in the case of large cities. The urban-scale polluting effect superimposing on the regional background pollution should not exceed short- and long-term air-quality standards, based on health and economic considerations.

Regional-scale pollution originates from point, area, and composite high sources (urban plumes). The length scale of regional pollution processes has been the topic of many discussions in recent years.

Smith (1973) suggested that the horizontal dimensions of an area which we call a region may lie between limits set by two meteorological properties. The lower limit is given by the horizontal distance beyond which the ground-level concentrations are significantly affected by the depth of the mixing layer. During the daytime this is typically of the order of 20 km downwind from the source area. The upper limit is given by the typical length over which meteorological parameters are relatively uniform. Bingemer (1977) found that the polluting effects (creating sulfur dioxide and particulate sulfate) of intense source areas could not be separated from the continental-background pollution beyond a downwind distance of 200–300 km.

It is generally agreed that the duration and distance within which the integrity of a polluted airmass is conserved depend on its initial volume and on the prevailing meteorological conditions. This is generally valid for urban and power-plant plumes on a regional scale. Therefore, it seems reasonable to use the length scale from 30–200 or 300 km for regional studies, depending on the homogeneity of the area investigated. The regional nature of many air-pollution problems (episodes and photochemical smog occurring at the regional level) generally indicates that the problem should be attacked at that level; this principle is recognized in the designation of official air-quality control regions in some countries. Air-quality standards on the regional level must be established, therefore, to avoid excess abatement costs in neighboring areas.

Nation-wide considerations could also play an important part in the overall air-quality management program, although national boundaries do not always correspond to air-pollution patterns (Rossano and Thielke, 1976).

Continental-scale considerations are important where air pollution may be transported from one country to another and create problems of international proportions. Continental-scale pollution originates from regional-level composite high sources (regional plumes), the horizontal scale of the processes being from 200 to 3000 km. The polluting effect of continental-scale processes superimposes on the global-background pollution, which includes mostly natural and long-lived anthropogenic components.

Concentrations of anthropogenic trace constituents are considered allowable as long as they do not cause detrimental acidification of soil and freshwater ecosystems over a considerable part of the continent. In continental-scale studies the role of the natural and long-lived components (originating from other continents) of global-background trace constituents needs further clarification.

Global-scale pollution originates from continental-scale composite high sources (continental plumes). The horizontal scale of global pollution processes is in excess of 3000 km. Global considerations include such factors as the balance between sources and sinks and evaluation of the effects that might possibly result from imbalance. It is necessary to know whether the global level of pollutants in the Earth's atmosphere is increasing, and if so, at what rate.

The proposed new system for scaling air-pollution processes helps in generalizing the concept of background and base-level pollution. Air pollution originating from a larger-scale pollution process around or outside a more-intense but smaller-scale process is called background pollution. The pollutant concentrations originating from the larger-scale process are superimposed on the more-intense effect of the smaller-scale process. By following this principle global, continental, regional, and urban background pollution can be defined (see Figure 2).

The mean value of the background pollution measured or estimated in synoptic situations when the more-intense but smaller-scale process causes the maximum polluting effect is called base-level pollution.

5 AIR-QUALITY MANAGEMENT

As mentioned earlier, the main goal of defining the new system for scaling air-pollution processes is to establish a practical means for simplifying the management of air quality, i.e. to create a general concept of air-quality management on revised scales. For the rational management of natural clean air, knowledge of the factors characterizing the emission sources, transmission, and air quality is necessary in sufficient spatial and temporal detail for the scale of the pollution process studied.

5.1 Sources of Trace Constituents

On a global scale most of the main trace constituents in the atmosphere originate from natural sources. The anthropogeneous contribution is considerable only in Europe and in the northeastern United States. The main types of anthropogenic emitters are point sources, area sources, and composite high sources.

Point sources are pollution emitters, whose plumes are not influenced (downwashed) by mechanical turbulence induced by surrounding buildings; therefore the natural diluting power of the atmosphere can be effective for the greater part of the year. High chimneys of power plants, district heating stations, and industrial plants, for example, are considered as point sources.

Area sources are those which emit through low chimneys or stacks, generally in the vicinity of the roof level of the surrounding buildings. Pollutants originating from an area source are subjected to mechanical turbulence induced by the surrounding buildings and thus an initial intense mixing occurs following their discharge. Industrial plants and city blocks, for example, are considered as area sources, their ventilation openings, low stacks, and chimneys being too numerous to be treated separately.

To study larger-scale pollution processes it is necessary to extend the concepts of point and area sources. The definition of regional-, continental-, and global-scale pollution processes is possible only by introducing the term "composite high source".

The composite high source is a generalized type of high point and area source. The horizontal dimension, Δy_j, of a composite high source of order j is defined by the cross-wind extent of the individual sources of order $(j-1)$ which make it up. The height, H_j, of the composite high source of order j is defined as being equal to the average mixing height, Z_j, characteristic of the pollution process of order $(j-1)$. The composite high source can be represented by a unit volume of air (Z m^3) having a pollutant concentration similar to the average air quality for a relatively homogeneous area of diameter Δy.

For convenience, the following j values are attributed to the pollution processes of various scales: local $j = 1$, urban $j = 2$, regional $j = 3$, continental $j = 4$, and global $j = 5$. For the sake of obtaining a uniform interpretation of the various kinds of pollution processes, it becomes necessary to introduce the concept of generalized total emission, $E_j(t)$. This is defined as the amount of nondecayed pollutants emitted during unit time by sources of order $(j-1)$ into the mixing layer of process j.

Therefore, the generalized total emission, $E_j(t)$ g/s, is given by the expression

$$E_j(t) = \sum_{i=1}^{n} E_{j\text{-}1,i}(t_0) F_i(P,S,T,R,X) \Delta t_i$$

where $E_{j\text{-}1,i}(t_0)$ is the emission of the ith polluting source in the pollution process of order j at time t_0, expressed in g/s; E_{0i} is the ith emission of local sources; $F_i(P,S,T,R,X)$ is the factor of decay of pollution (in percent/unit time) depending on the rate and amount of precipitation, intensity of solar radiation, temperature, humidity, and concentration of other pollutants; and $\Delta t_i = t_i - t_0$ is the transmission time for air pollution originating from the $(j-1)$th source and arriving at the boundary of process of order j.

In other words, the pollution processes on various scales are interconnected in a cumulative way. For pollutants with short residence times it is most important that the decay factor is taken into account properly. The average density of emission is scale dependent, its maximum being on the local scale and its minimum on the global scale.

5.2 Transmission of Air Pollutants

Pollutants emitted into the atmosphere are influenced by environmental factors such as wind, precipitation, solar radiation, temperature, humidity, other gaseous pollutants, and aerosols. Under the influence of these factors their concentration and residence times in the atmosphere vary considerably.

The simultaneous effects of transport, dispersion, removal, and transformation mechanisms on pollutants are referred to as the "transmission" of air pollutants. The priorities of the main factors of transmission for various scales of pollution processes are shown in Table 2. It can be concluded that, as the scales of pollution processes vary, the relative importance of factors of transmission changes:

TABLE 2 Priorities of the main factors of transmission for different-scale pollution processes (D = wind direction, U = wind speed, P = wind pattern (at the height of the plume), T = trajectory (925 mbar), τ_d = dry deposition (wind speed, surface roughness), τ_w = wet deposition (duration and intensity of precipitation), τ = transformation (solar radiation, temperature, relative humidity), σ = dispersion, Z = mixing height, and H = effective height of point and area sources).

Priorities	Local 0–10 (20) km	Urban 0–30 (100) km	Regional 30–200 (300) km	Continental 200–3000 km	Global > 3000 km
1	D	P	P	T	τ_w
2	H_p	U	Z	τ_w	τ
3	U	τ_d	τ_w	Z	T
4	σ	Z	τ_d	U	–
5	Z	τ	τ	τ_d	–
6	τ_d	τ_w	U	τ	τ_d
7	τ_w	σ	H_p	–	--
8	τ	H_a	σ	–	U

(a) Mixing depths, relatively unimportant for urban models, become increasingly important in evaluating the growth of an urban plume.

(b) Tall stacks, of great use in controlling urban air pollution, are much less effective in altering the air quality on regional scales of 30 to 200 km.

(c) Land-use planning, of limited effectiveness in controlling the transport of pollutants on an urban scale, is of vastly increased importance in regional urban planning.

(d) For a large source, the maximum ground-level concentration near that source is the limiting consideration. For regional air-quality planning the cumulative effect of large urban sources can be pronounced if they are aligned parallel to regional wind trajectories.

(e) The effects of turbulent dispersion should be taken into account for local-, urban-, and regional-scale studies, but are unimportant for continental- and global-scale work.

(f) As wind pattern has first priority on most scales, it is recommended that the effects of climatic fluctuations on air- and precipitation-quality trends are studied and clarified.

5.3 Air Quality

Air-quality aspects of air-pollution processes are characterized by the maximum polluting effect (maximum concentration), the background (base-level) pollution, and the prevailing air-quality standards. The main object of interest is usually an estimate of the maximum polluting effect of a source or pollution process which may develop during certain meteorological conditions. The polluting effect of the process superimposes on the background pollution originating from a higher-order pollution process.

From the viewpoint of air-quality management it is important to distinguish between background and base-level pollution. The main difference is that the base-level pollution, X_b, is the average value of background pollution originating from the jth-order pollution process, measured or estimated from concentration values formed in synoptic situations in which the polluting process of order $j - 1$ exerts its maximum polluting effect, $X_{p\,max}$.

From this it is evident that the maximum pollution, X_{max}, will be equal to

$$X_{max} = X_b + X_{p\ max}$$

The main characteristics of air-quality measurements depend on the scale of the pollution process. The respective siting criteria, number of stations, and period of investigation are shown by Table 3. On this basis, it can be concluded that the effects of an air-pollution process can best be measured at a receptor point if smaller-scale processes are absent for a distance d around the point (see Table 3). The representativeness of the receptor point can be checked by correlating the measured and calculated concentration values originating from different-scale pollution processes. The location of the receptor has been properly selected if the highest correlation results.

TABLE 3 Air-quality measurements and standards.

Characteristics	Local	Urban	Regional	Continental	Global
Distance upwind from receptor (km)	0–10 (20)	0–30 (100)	30–200 (300)	200–3000	> 3000
Siting criteria (km)	distance of local sources $d = X_{max}$	$d > 0.1$	$d > 40–60$	$d > 100$	$d > 500$
Number of stations		5–50/city	100	50–80/ continent	8–10
Period of investigation	few days	1–5 years	cont. d	3–5 years	cont. d
	Mainly based on				
			Avoidance of		
				Detrimental effects	
Air-quality standard	Health and economic considerations		Excess abatement costs in other areas	On fauna and flora	On climate

As for air-quality standards which are valid for local- and urban-scale pollution processes, air- and precipitation-quality criteria and standards should also be established for regional-, continental-, and global-scale pollution processes. Some initial objectives are shown in Table 3.

The loading capacity of the available air resource can be expressed by using the equation

$$X_{n\ max} > X_{max} = X_b + X_{p\ max}$$

where $X_{n\,max}$ is the threshold value of air- or precipitation-quality standards. This means that air-quality management is possible within the limits given by an air- or precipitation-quality standard, $X_{n\,max}$, and base-level pollution, X_b

$$X_{n\,max} - X_b > X_{p\,max}$$

REFERENCES

Bingemer, H. (1977). Transport und Abbau schwefelhaltiger Luftverunreinigungen im Lee grosser Flächen-quellen. Diplomarbeit, Universität Frankfurt am Main.

Flohn, H. (1959). Bemerkungen zum Problem der globalen Klimaschwankungen. Arch. Met. Geophys. Bioklim., B9:1–13.

Georgii, H.W. (1978). Large scale spatial and temporal distribution of sulfur compounds. Atmos. Environ., 12:681–690.

Hall, W.F. and Holloway, L. (1955). System analysis of data provision and processing. Progress Rep. No. 2, Contract No. CSO and A54-24.

Hidy, G.M., Mueller, P.K., and Tong, E.Y. (1978). Spatial and temporal distributions of airborne sulfate in parts of the United States. Atmos. Environ., 12:735–752.

Mason, B.J. (1970). Future developments in meteorology: an outlook to the year 2000. Q. J. R. Meteorol. Soc., 96:349–368.

Mészáros, E. (1977). A levegőkémia alapjai. Akadémiai Kiadó, Budapest, p. 180.

Mészáros, E. (1978). Concentration of sulfur compounds in remote continental and oceanic areas. Atmos. Environ., 12:699–705.

Ottar, B. (1978). An assessment of the OECD study on long range transport of air pollutants (LRTAP). Atmos. Environ., 12:445–454.

Perhac, R.M. (1978). Sulfate regional experiment in NE United States: the SURE program. Atmos. Environ., 12:641–647.

Rossano, A.T. and Thielke, J.F. (1976). The design and operation of air quality surveillance systems. Manual on urban air quality management. WHO Regional Publications, European Series No. 1, 153–177.

Smith, F.B. (1973). Discussion on the role of regional air pollution studies in model development. CCMS No. 30.

Szepesi, D. (1974). Factors and concepts of air resource management. Időjárás, 78:325–332.

Yoshino, M.M. (1961). Microclimate, an Introduction to Local Meteorology. Chijinshokau, Tokyo.

Yoshino, M.M. (1975). Climate in a Small Area. University of Tokyo Press, Tokyo.

Takahashi, K. (1969). Synoptic Meteorology. Iwanami, Tokyo.

WMO (1976). Report on Expert Meeting on Siting Criteria. WMO, Mainz, 26–28 October 1976.

WMO (1978). Report on Siting Criteria Review Meeting. WMO, Geneva, 28–31 March 1978.

Part Two

Models for Real-Time Prediction and Control of Air Pollution

SHORT-TERM FORECASTING OF LOCAL WINDS BY BLACK-BOX MODELS

C. Bonivento and A. Tonielli
Istituto Automatica, Università di Bologna, Bologna (Italy)

1 INTRODUCTION

A wind forecast is a relevant input to real-time pollution predictors, i.e. to recursive algorithms, which at the beginning of each time step supply the "most likely" values of future pollutant concentration on the basis of current meteorological and concentration measurements (see for instance Fronza et al., 1982).

In this paper we illustrate the forecasting performance of "black-box" hourly wind predictors as recorded in two case studies (of the regions of Sassuolo and Venice in Northern Italy). More specifically, we describe three types of wind predictor derived from the following stochastic models of the AutoRegressive Moving Average (ARMA) type (see for instance Box and Jenkins, 1970).

(i) An AutoRegressive (AR) model of wind speed. In this mathematical representation the average wind *speed* in each hour is expressed as a linear combination of previous hourly wind-speed values plus white noise (a purely random term).

(ii) A bivariate AR model of the wind components (westerly and southerly) in the horizontal plane. In this model each hourly wind *component* is expressed as a linear combination of previous values of both hourly components plus white noise.

(iii) A bivariate ARMA model (a model with a colored noise term—specifically, with a moving average noise).

The performance of the three predictors in the two case studies leads to the following conclusions.

(1) The use of the bivariate model (ii) yields a significant improvement in wind forecasting compared with approach (i). This means that cross-correlations between the two wind components are useful information to take into account.

(2) The introduction of colored noise, i.e. the use of model (iii) instead of (ii), does not give a significantly better forecast.

2 THE STOCHASTIC MODELS AND PREDICTORS

2.1 The Univariate AR Model and Predictor

The record of hourly wind speed was first taken into account. In both cases this record showed a strong daily periodicity (mainly due to breeze phenomena) so that the data were cyclically standardized ($\Delta = 1$ h) as follows

$$w(24k + r) = [w(24k + r) - \mu_r]/\sigma_r \tag{1}$$

where $w(24k + r)$ is the hourly wind speed in the rth hour ($r = 1,2,\ldots,24$) of the kth day, i.e. in the interval $[(r-1)\Delta, r\Delta]$ of the kth day and μ_r and σ_r are the mean and the standard deviations of the wind speed in the rth hour of the day. The stochastic process $\{w(t)\}_t$ was then described by the AR model

$$w(t+1) = \sum_{j=1}^{p} a_j w(t-j+1) + n(t+1) \tag{2}$$

where $\{n(t)\}_t$ is the white noise and the a_j are the model parameters.

The parameters of model (2) were estimated by a standard least-squares fitting technique.

The real-time predictor derived from eqn. (2) is given (see for instance Box and Jenkins, 1970) by

$$\hat{w}(t+1|t) = a_1 w(t) + a_2 w(t-1) + a_3 w(t-2) \tag{3a}$$

$$\hat{w}(t+2|t) = a_1 \hat{w}(t+1|t) + a_2 w(t) + a_3 w(t-1) \tag{3b}$$

$$\hat{w}(t+3|t) = a_1 \hat{w}(t+2|t) + a_2 \hat{w}(t+1|t) + a_3 w(t) \tag{3c}$$

$$\hat{w}(t+4|t) = a_1 \hat{w}(t+3|t) + a_2 \hat{w}(t+2|t) + a_3 \hat{w}(t+1|t) \tag{3d}$$

where (for $f = 1,2,3,4$) $\hat{w}(t+f|t)$ is the forecast of $w(t+f)$ made at the instant $t\Delta$. Of course, if $t+f$ corresponds to the rth hour of the day the forecast of the variable of interest is found [see eqn. (1)] to be

$$\hat{w}(t+f|t) = \mu_r + \sigma_r \hat{w}(t+f|t) \tag{4}$$

2.2 The Bivariate AR Model and Predictor

First let $w_1(t)$ be the tth (cyclically standardized) hourly westerly component of the wind, $w_2(t)$ the tth (cyclically standardized) hourly southerly component of the wind, and

$$w(t) = |w_1(t)\ w_2(t)|^{\mathrm{T}}$$

where the superscript T is the vector transposition symbol. Consider the following bivariate AR model of the wind-vector stochastic process $[w(t)]_t$

$$w(t+1) = \sum_{j=1}^{p} b_j w(t-j+1) + m(t) \tag{5}$$

where $\{m(t)\}_t$ is the white noise and the b_j are the model parameters.

The parameters of model (5) were also estimated by standard least-squares fitting. The predictor derived from eqn. (5) is quite similar to that derived from eqns. (3) and (4) except that in this case the forecast is a vector.

2.3 The Bivariate ARMA Model and Predictor

The following ARMA model of the wind vector was considered

$$\begin{vmatrix} f(z) - g_{11}(z) & -g_{12}(z) \\ -g_{21}(z) & f(z) - g_{22}(z) \end{vmatrix} w(k) = f(z)\, \boldsymbol{v}(k) \tag{6}$$

where z denotes the forward shift operator in the time domain,

$$f(z) = z^n + a_1 z^{n-1} + \ldots + a_n$$

$$g_{sq}(z) = g_{sq,1} z^{n-1} + g_{sq,2} z^{n-2} + \ldots + g_{sq,n} \quad s = 1,2; q = 1,2$$

and $\{\boldsymbol{v}(k)\}_k$ is the zero-mean white noise.

Model (6) can be considered as an input–output relationship corresponding to the steady-state innovation filter representation (see for instance Goodwin and Payne, 1977)

$$\hat{x}(k+1|k) = F\hat{x}(k|k-1) + K\boldsymbol{v}(k) \tag{7a}$$

$$w(k) = H\hat{x}(k|k-1) + \boldsymbol{v}(k) \tag{7b}$$

where K is the steady-state Kalman gain and $\{\boldsymbol{v}(k)\}_k$ is the innovation sequence.

Precisely, by letting

$$G_m = \begin{vmatrix} g_{11,m} & g_{12,m} \\ g_{21,m} & g_{22,m} \end{vmatrix} \qquad m = 1,2, \ldots, n$$

we obtain the following relationships between the parameters of the two representations (6) and (7) (see for instance Spain, 1971):

$f(z)$ is the minimal polynomial of F (8)

$$G_m = \sum_{k=0}^{m-1} a_{m-k-1} H(F-KH)^k K \quad (a_0 = 1) \tag{9}$$

From estimates of $[a_m, G_m]_{m=1}^n$ and through eqns. (8) and (9) it is possible to determine an innovation filter state representation, i.e. a triplet F, K, H (see for instance Tajima, 1978). However, if one is only interested in the forecast $\hat{w}(k+1|k) = H\hat{x}(k+1|k)$ of the variable of interest (see also Aasnaes and Kailath, 1973) it is straightforward to derive a prediction formula which makes use only of the estimates of $[a_m, G_m]_{m=1}^n$. In fact, from eqn. (7b) the formula is found to be

$$\nu(k) = w(k) - \hat{w}(k|k-1)$$

and hence by substitution into eqn. (6) the "canonical" predictor is

$$f(z)\hat{w}(k|k-1) = G(z)w(k) \tag{10}$$

where

$$G(z) = \begin{vmatrix} g_{11}(z) & g_{12}(z) \\ g_{21}(z) & g_{22}(z) \end{vmatrix}$$

The explicit form of the one-step-ahead predictor (10) is

$$\hat{w}(k+n|k+n-1) = -\sum_{m=1}^{n} a_m \hat{w}(k+n-m|k+n-m-1) + \sum_{m=1}^{n} G_m w(k+n-m)$$

The parameters $[a_m, G_m]_{m-1}^n$ were estimated via the maximum-probability technique, i.e. the maximum of the probability function was sought using Rosenbrock's algorithm (Rosenbrock, 1960).

3 DISCUSSION OF THE RESULTS

3.1 The Sassuolo Case

The performance of the three predictors was tested on hourly wind data recorded at the Sassuolo meteorological station in a spring month in 1974. The data for another month were used for estimating the parameters and orders of the various models. The results are summarized in Tables 1–5 for various forecasting steps (in the tables, ρ is the

TABLE 1 The forecasting performance (for wind speed) of the univariate AR predictor at Sassuolo.

Forecasting step	ρ	σ_e
1	0.8663	0.775
2	0.7357	1.053
3	0.6345	1.206
4	0.5601	1.299

TABLE 2 The forecasting performance (for wind components) of the bivariate AR predictor at Sassuolo.

Forecasting step	Westerly component		Southerly component	
	ρ	σ_e	ρ	σ_e
1	0.9517	0.974	0.9033	0.709
2	0.8958	1.409	0.8152	0.957
3	0.8644	1.591	0.7368	1.117
4	0.8424	1.703	0.6700	1.226

TABLE 3 The forecasting performance (for wind speed and direction) of the bivariate AR predictor at Sassuolo.

Forecasting step	Wind speed		Wind direction	
	ρ	σ_e	ρ	σ_e
1	0.8688	0.859	0.8740	1.089
2	0.7463	1.196	0.7954	1.373
3	0.6641	1.365	0.7258	1.634
4	0.6015	1.482	0.6663	1.832

TABLE 4 The forecasting performance (for wind components) of the bivariate ARMA predictor at Sassuolo.

Forecasting step	Westerly component		Southerly component	
	ρ	σ_e	ρ	σ_e
1	0.9517	0.974	0.9038	0.707
2	0.8958	1.408	0.8158	0.956
3	0.8643	1.592	0.7373	1.116
4	0.8424	1.703	0.6711	1.225

TABLE 5 The forecasting performance (for wind speed and direction) of the bivariate ARMA predictor at Sassuolo.

Forecasting step	Wind speed		Wind direction	
	ρ	σ_e	ρ	σ_e
1	0.8689	0.859	0.8669	1.113
2	0.7477	1.194	0.7921	1.383
3	0.6648	1.366	0.7254	1.635
4	0.6020	1.483	0.6658	1.832

TABLE 6 The forecasting performance (for wind components) of the persistent wind-speed predictor at Sassuolo.

Forecasting step	Westerly component		Southerly component	
	ρ	σ_e	ρ	σ_e
1	0.9178	1.284	0.8831	0.798
2	0.7782	2.109	0.7638	1.135
3	0.6371	2.696	0.6505	1.380
4	0.4917	3.189	0.5451	1.575

correlation between the predictions and observations and σ_e is the standard deviation of the forecasting error). We can draw the following basic conclusions:

(a) There is a significant improvement in moving from the univariate AR model to the bivariate AR model (see Tables 1 and 3); i.e. the cross-correlation between the wind components is "valuable information".

(b) There is no relevant improvement in moving from the bivariate AR model to the bivariate ARMA model (Tables 3 and 5).

(c) The performance is, in general significantly better than that of the trivial persistence predictor (see Tables 1, 3, and 6).

3.2 The Venetian Case

The same conclusions can be drawn by analyzing the results for the Venetian case which for brevity are summarized in a less-expanded form in Tables 7–9. The univariate AR model has been used as input to the SO_2 concentration predictor (see Fronza et al., 1982), and the bivariate AR model will be applied in the near future.

TABLE 7 The forecasting performance (for wind speed) of the univariate AR predictor at Venice.

Forecasting step	ρ
1	0.81
2	0.64
3	0.55
4	0.52

TABLE 8 The forecasting performance (for wind components) of the bivariate AR predictor at Venice.

Forecasting step	ρ	
	Westerly component	Southerly component
1	0.80	0.92
2	0.64	0.86
3	0.56	0.84
4	0.53	0.82

TABLE 9 The forecasting performance (for wind components) of the persistent wind-speed predictor at Venice.

Forecasting step	ρ	
	Westerly component	Southerly component
1	0.76	0.89
2	0.52	0.76
3	0.36	0.62
4	0.24	0.46

REFERENCES

Aasnaes, H.B. and Kailath, T. (1973). An innovations approach to least-squares estimation. Part VII: some applications of vector autoregressive moving average models. IEEE Trans. Autom. Control, 18:601–607.

Box, G.E.P. and Jenkins, G.M. (1970). Time Series Analysis, Forecasting and Control. Holden–Day, San Francisco, California.

Fronza, G., Spirito, A., and Tonielli, A. (1982). Kalman prediction of sulfur dioxide episodes. In G. Fronza and P. Melli (Editors), Mathematical Models for Planning and Controlling Air Quality. Pergamon Press, Oxford. (This volume, p. 161.)

Goodwin, G.C. and Payne, R.L. (1977). Dynamic System Identification: Experiment Design and Data Analysis. Academic Press, New York.

Rosenbrock, H.H. (1960). An automatic method for finding the greatest or least value of a function. Comput. J., 3:175–184.

Spain, D.S. (1971). Identification and modeling of discrete, stochastic, linear systems. Tech. Rep. 6302-10. Information Systems Laboratory, Stanford University, Stanford, California.

Tajima, K. (1978). Estimation of steady-state Kalman filter gain, IEEE Trans. Autom. Control, 23: 944–945.

A K MODEL FOR SIMULATING THE DISPERSION OF SULFUR DIOXIDE IN AN AIRSHED

E. Runca and P. Melli
Centro Scientifico IBM, Rome (Italy)

A. Spirito
Centro Teoria dei Sistemi, Milan (Italy)

1 INTRODUCTION

The need to obtain a detailed description of both the spatial and the temporal evolution of pollutant concentration in complex urban and/or industrial areas led as early as 1970 (Randerson, 1970) to consideration of the possibility of developing three-dimensional models based on the integration of the atmospheric-diffusion equation. Later a more systematic and complex model was developed by Shir and Shieh (1974) who applied it to the St. Louis, Missouri area. An application of the same model was developed some years later to simulate the hourly SO_2 concentration in the Venetian Lagoon area (Marziano et al., 1979).

In 1974 the idea of using a numerical model for real-time estimation of air pollution was proposed by Desalu et al. (1974) but to our knowledge the first application of this idea to a multiple-source case is due to Bankoff and Hanzevack (1974). The model described in the present paper was first developed in a rather general form (Runca, 1976; Sardei and Runca, 1976) and was afterwards modified for use in a real-time predictor which was then applied in the forecasting of air-pollution episodes in the Venice region.

2 THE ATMOSPHERIC-DIFFUSION EQUATION

Starting from the continuity equation for an air pollutant, the classical atmospheric-diffusion equation

$$(\partial C/\partial t) + V \cdot \nabla C = \nabla \cdot (\bar{K} \cdot \nabla C) + S + R \tag{1}$$

can be derived under the assumptions that the gradient-transfer theory holds for atmospheric turbulent-diffusion processes, that the wind field is nondivergent, and that molecular diffusion is negligible in comparison with turbulent diffusion. In eqn. (1) C and V are the pollutant concentration and the wind field respectively, $\bar{K}$ is the eddy-diffusivity tensor, and S and R are source and removal terms, respectively.

It should be pointed out that the concentration and the wind field in eqn. (1) are averages over a time interval which is large compared with the dominant time scale of turbulent fluctuations but small in comparison with the time scale of variations of the mean concentration and the wind speed.

3 THE NUMERICAL INTEGRATION SCHEME

The numerical problems in integrating eqn. (1) have been amply discussed in the literature. The main difficulty stems from the inability of conventional finite-difference schemes to describe accurately the advection terms in eqn. (1). In fact, such schemes do not move air particles along the wind trajectories (i.e. the travel distance per time step is not equal to the mesh spacing); this produces both downwind and upwind an "artificial numerical diffusion" which can be of the same order of magnitude as the computed quantity (see for example Roberts and Weiss, 1966).

In order to reduce artificial diffusion, several algorithms have been proposed: they range from the method of Egan and Mahoney (1972) involving the use of the first three moments of concentration distribution in each grid element to mixed Eulerian–Lagrangian schemes such as the "particle-in-a-cell" method (Sklarew et al., 1971; Lange, 1973), to the method proposed by Runca and Sardei (1975), and finally to the use of Galerkin techniques (Christensen and Prahm, 1976; Melli, 1976). All these methods, however, have at least some unsatisfactory features, such as, depending on the particular method, severe stability constraints, difficulty in treating boundary conditions, and computer programming complexity. These drawbacks are largely avoided in the scheme described in the following, which allows for a nonuniform grid in all directions, requires a reasonably short amount of computer time, can be programmed simply, and gives satisfactory accuracy. This scheme is based on the fractional-step algorithm (see for example Yanenko, 1971) and uses Carlson's method for treating the advection terms and the Crank–Nicolson method for the diffusive terms. On assuming that the vertical component of the wind speed is negligible and on neglecting the removal terms, eqn. (1) reduces to

$$
\begin{aligned}
&(\partial C/\partial t) + v_x[z,s(t)]\,\partial C/\partial x + v_y[z,s(t)](\partial C/\partial y) \\
&= K_x[s(t)](\partial^2 C/\partial x^2) + K_y[s(t)](\partial^2 C/\partial y^2) \\
&\quad + \partial\{K_z[z,s(t)](\partial C/\partial z)\}/\partial z + S(x,y,z,t) \qquad (2)
\end{aligned}
$$

with the initial and boundary conditions

$$C(x,y,z,0) = 0 \qquad (3)$$

$$C(x_W,y,z,t) = C(x_E,y,z,t) = 0 \qquad \forall y,z;\forall t \qquad (4)$$

$$C(x,y_S,z,t) = C(x,y_N,z,t) = 0 \qquad \forall x,z;\forall t \tag{5}$$

$$K_z(\partial C/\partial z) = 0 \quad z = 0,H \qquad \forall x,y;\forall t \tag{6}$$

where H is the height of the inversion-layer base, v_x and v_y are the x and y wind components, K_x, K_y, and K_z are the x, y, and z diffusion coefficients, and x_W, x_E, y_N, and y_S are the abscissas of the western, eastern, northern, and southern boundaries of the integration region.

According to the method of fractional steps eqn. (2) is first split into the following six equations:

$$\partial C/\partial t = \sum_{r=1}^{R} Q_r \delta(z-z_r)\delta(y-y_r)\delta(x-x_r) \tag{7}$$

$$\partial C/\partial t = -v_x[z,s(t)]\,\partial C/\partial x \tag{8}$$

$$\partial C/\partial t = -v_y[z,s(t)]\,\partial C/\partial y \tag{9}$$

$$\partial C/\partial t = K_x[s(t)]\,\partial^2 C/\partial x^2 \tag{10}$$

$$\partial C/\partial t = K_y[s(t)]\,\partial^2 C/\partial y^2 \tag{11}$$

$$\partial C/\partial t = \partial\{K_z[z,s(t)]\,\partial C/\partial z\}/\partial z \tag{12}$$

where $\delta(\,)$ stands for Dirac's function, Q_r is the emission rate of the rth source and x_r, y_r, and z_r are the coordinates of the rth source. Equation (7) takes into account the contribution of the emission term, eqns. (8) and (9) the two advection terms, and eqns. (10)–(12) the contribution of the diffusion terms.

3.1 The Contribution of the Source Term

The solution of eqn. (7) consists of adding to the concentration field at time $k\Delta t$ the contribution of the source, which is distributed within the box in which the source is located. The finite-difference analog of eqn. (7) is found to be

$$C^*_{ijm} = C^k_{ijm} + \gamma_{ijm} Q_r \Delta t/\Delta x_r \Delta y_r \Delta z_r \tag{13}$$

where Δx_r, Δy_r, and Δz_r are the dimensions of the box containing the rth source, γ_{ijm} is a geometrical factor depending on source position, and C^* is the concentration field after the addition.

3.2 The Contribution of the Advection Terms

Equations (8) and (9) are integrated by using Carlson's scheme (Richtmyer and Morton, 1967). This scheme leads to the following two difference equations (for simplicity only the analogs corresponding to eqn. (8) are reported):

$$C_{ijm}^{**} = C_{ijm}^{*} - (v_{x_m} \Delta t/\Delta x_i)(C_{ijm}^{*} - C_{(i-1)jm}^{*}) \quad \text{if } v_{x_m} \Delta t/\Delta x_i \leqslant 1 \tag{14}$$

$$C_{ijm}^{*} = C_{(i-1)jm}^{*} - (\Delta x_i/v_{x_m} \Delta t)(C_{(i-1)jm}^{**} - C_{ijm}^{*}) \quad \text{if } v_{x_m} \Delta t/\Delta x_i \geqslant 1 \tag{15}$$

where C^{**} is the concentration field after the solution of eqn. (8). The situations represented by eqns. (14) and (15) are shown in Figure 1 by the trajectories I and II respectively in an (x, t) plane (j = constant, m = constant). This algorithm is unconditionally stable and allows the reduction of the numerical diffusion to zero at points where $v_{x_m} \Delta t = \Delta x$. Analogs similar to eqns. (14) and (15) are used for transport along y, thus providing a new concentration field C^{***}.

3.3 The Contribution of Diffusion Terms

Equations (10)–(12) are numerically approximated by the method due to Crank and Nicolson (1947):

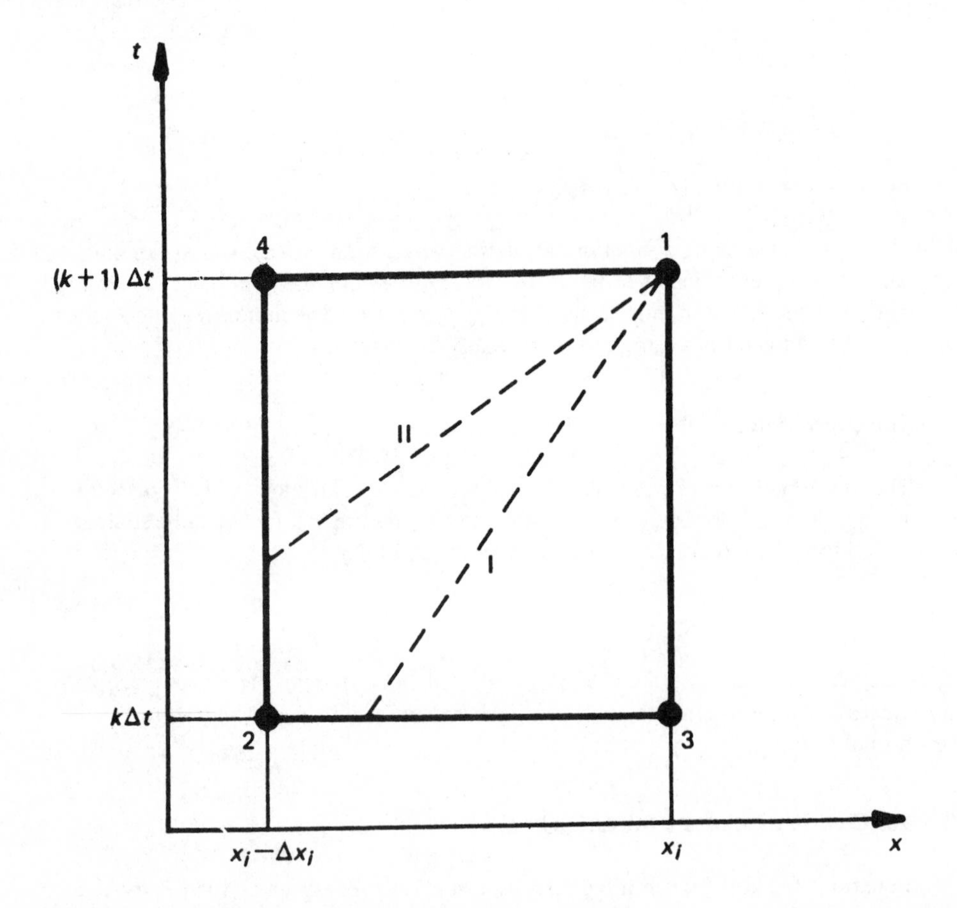

FIGURE 1 Carlson's scheme in the (x,t) plane: I is the trajectory in the case $v_{x_m} \Delta t < \Delta x_i$: II is the trajectory in the case $v_{x_m} \Delta t \geqslant \Delta x_i$.

$$C^{****}_{ijm} = C^{***}_{ijm} + \tfrac{1}{2}[D_x(C^{****}_{ijm}) + D_x(C^{***}_{ijm})] \quad (16)$$

where

$$D(C_{ijm}) = \frac{2\Delta t}{\Delta x_i(\Delta x_i + \Delta x_{i-1})} \times \left[\frac{\Delta x_i}{\Delta x_{i-1}} K_{x_{i-1/2}} C_{(i-1)jm} - \left(\frac{\Delta x_i}{\Delta x_{i-1}} K_{x_{i-1/2}} + K_{x_{i+1/2}}\right) C_{ijm} + K_{x_{i+1/2}} C_{(i+1)jm}\right] \quad (17)$$

is the standard centered difference operator for diffusion with variable grid-spacing and diffusion coefficients and C^{****} is the concentration field after taking account of diffusion in the x direction. The result of similar procedures applied to eqns. (11) and (12) is taken as an approximation of C^{k+1}, i.e. of the concentration field at time $(k + 1)\Delta t$.

The basic difference between the algorithm described here and other fractional-step schemes in the air-pollution literature (Shir and Shieh, 1974; Bankoff and Hanzevack, 1975) is that the latter treat only the vertical diffusion step by an implicit formulation. The explicit treatment of the other terms causes stability limitations which may increase the computational burden. In contrast, the algorithm adopted in the present study, is unconditionally stable and therefore there are no limitations as to the definition of horizontal grid spacings. This results in the advantage of being able to use smaller grid spacings where large nonuniformities occur (e.g. in industrial areas where most important sources are located) and larger spacings where concentration gradients are smaller.

Moreover a horizontally nonuniform grid allows the use of monitoring stations coincident with grid points, thus avoiding any interpolation in the comparison of forecast and observed data.

4 THE APPLICATION TO THE VENETIAN CASE

4.1 Description of the Area

The area considered (Figure 2) is located in the northeastern part of Italy and includes the urban centers of Venice, Marghera, and Mestre, and the large industrial area of Porto Marghera. An analysis of the occurrence of episodes in the whole area (Zannetti et al., 1977) has shown that most episodes occur in the industrial area; the few episodes in Venice are recorded in conjunction with well-defined meteorological conditions (either light winds blowing from north-northwest or stagnant conditions in the whole region) and are less intense. Hence the use of concentration measurements in the historical center would not add much information to an air-quality control scheme, which was the ultimate aim of the work. Thus the present model was implemented only for the whole region shown in Figure 2. Of course, this choice significantly reduced the computational effort required.

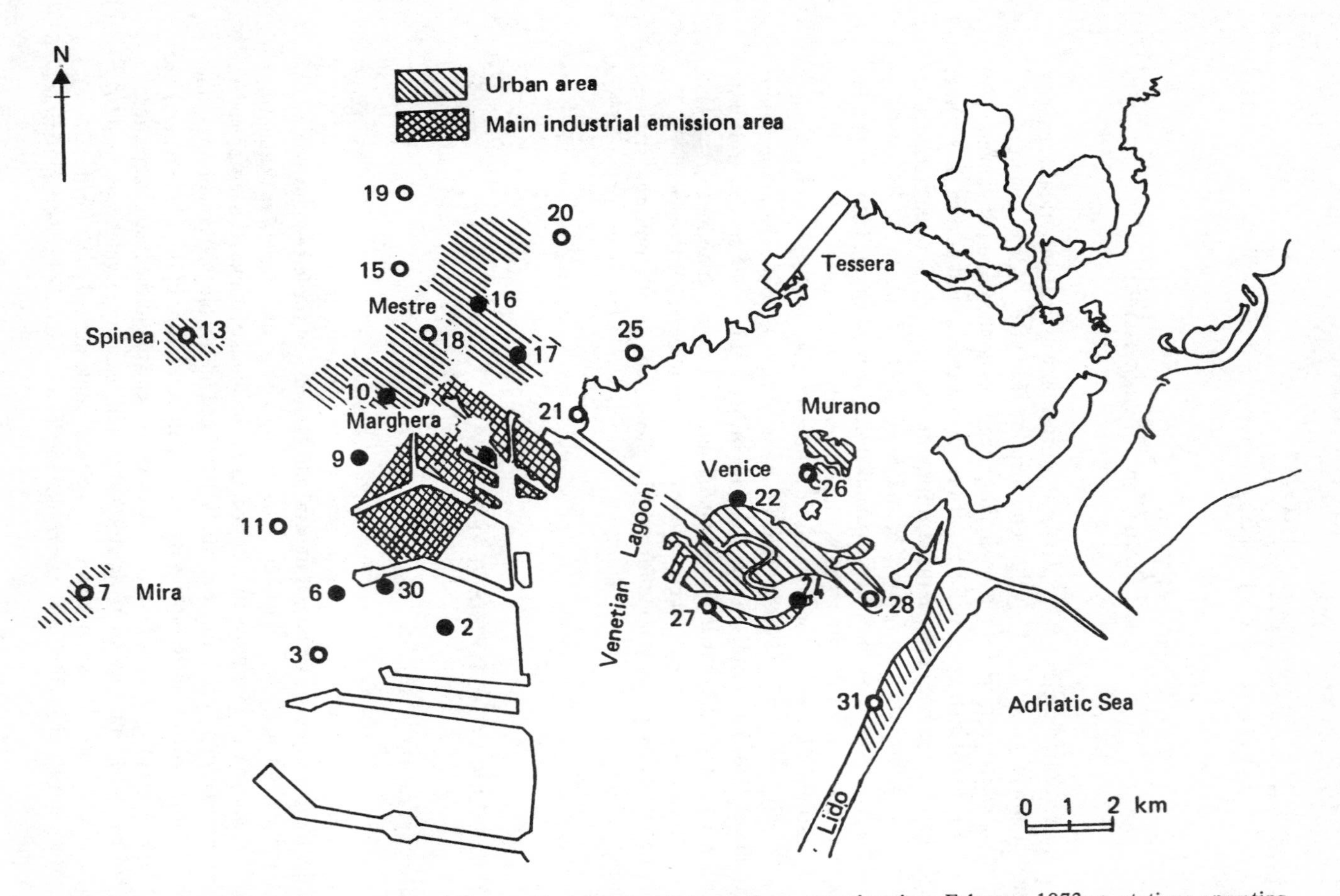

FIGURE 2 The Venetian lagoon area and the monitoring network (•, stations operating since February 1973; ○, stations operating since February 1974).

4.2 The Input Data Set

The available emission, meteorological and SO_2-concentration data were as follows.

Data for each of the 74 industrial sources (distributed in the industrial area shown in Figure 2) were directly obtained from 1971 National Census figures. To give an idea of the overall emission in the region, the estimated pollutant released from industries amounts to about 160,000 tons per year, in addition to approximately 10,000 tons per year due to domestic heating. The location and average SO_2 emission rate of each source were available. Plume rises were computed using the CONCAWE formula. Both the meteorological and the concentration data used in the present study were obtained from the monitoring network (see Figure 2) installed by Tecneco on behalf of the Governmental Department of Health. The network consists of one meteorological station and 24 SO_2-monitoring sensors. The meteorological station is 15 m above ground level and records hourly wind speed and direction, temperature, pressure, humidity, rainfall, cloudiness, and fog. The wind direction is recorded according to the eight sectors of the compass, thus introducing an indeterminacy of $\pm 22.50°$. The concentration data recorded by the 24 monitoring sensors are transmitted to a small computer which elaborates the data and records the hourly average values as well as daily statistics. In 1973, the year to which this study refers, only ten stations were in operation. Two of these are located in Venice itself; the rest are the stations considered in the present application. In general the concentration data exhibit satisfactory reliability but this is not true of the other types of data. In fact the emission data are only average rates; this is a very rough input in modeling an episode partly or mainly due to extra release. As for the meteorological measures, one station alone obviously cannot record the spatial variation of the wind and diffusion parameters.

4.3 Model Specifications

The region of interest (16.5 km $\times$ 18.0 km) was discretized by means of $10 \times 12 \times 7$ grid points. The horizontal grid spacings range from a minimum of 1 km to a maximum of 2.5 km (see Figure 3 where the monitoring stations, which all coincide with grid points, are also shown). The vertical grid sizes were specified as follows:

$$\Delta z_m = \begin{cases} 50 \text{ m} & m = 1,2 \\ 75 \text{ m} & m = 3,4 \\ (H - 250)/2 & m = 5,6 \end{cases}$$

Since no measurement of the mixing depth H was available, H was kept constant at 500 m in all simulations.

The degree of refinement of the chosen grid lies somewhere between that of the grid used by Shir and Shieh (they used $40 \times 30 \times 14$ points for a region 60 km $\times$ 45 km approximately) and that of the grid used by Bankoff and Hanzevack who discretized by means of $8 \times 16 \times 4$ points for a region 24 km $\times$ 32 km. The compromise was dictated by the need to keep within reasonable limits both the computing burden and the effort required to provide a satisfactory degree of resolution.

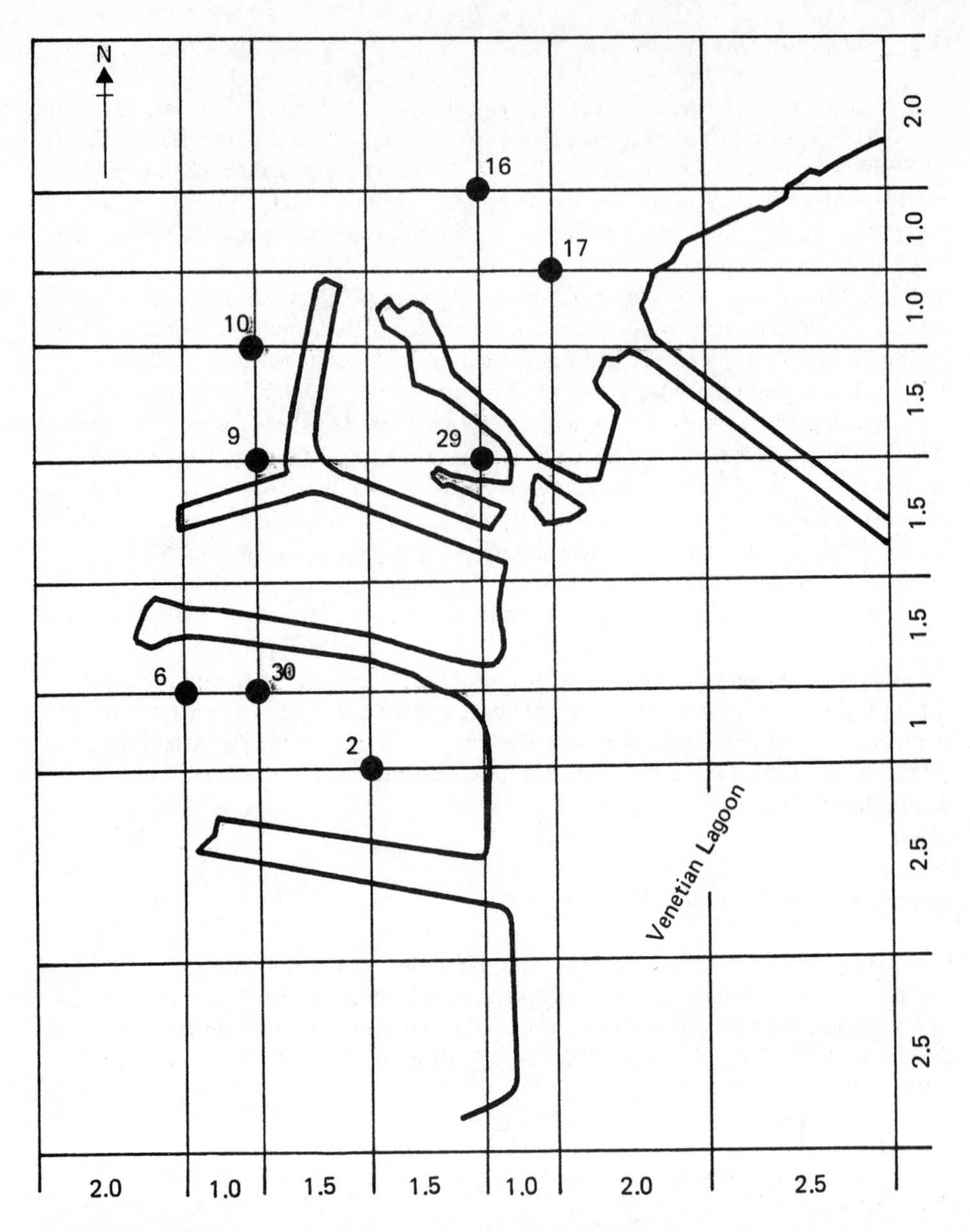

FIGURE 3 The geometry of the grid in the horizontal plane with distances in kilometers (•, stations working since February 1973, labeled in accordance with the Tecneco classification).

The atmospheric stability was classified according to Pasquill's (Pasquill, 1974) categories ranging from A (strong instability) to F (extreme stability) on the basis of wind and cloudiness data supplied by the meteorological station. Because of the lack of detailed information, the wind field was assumed to be constant over the whole area and values at upper levels were computed by means of the well-known power-law formula

$$v_m(t) = v_R(t)(z_m/z_R)^{\alpha[s(t)]} \tag{18}$$

where $v_m(t)$ is the wind vector at the mth level, $v_R(t)$ is the wind vector supplied by the meteorological station at level z_R (15 m), and $\alpha[s(t)]$ is the given function of stability (reported in Table 1).

TABLE 1 The wind and diffusion parameters versus Pasquill's stability classes.

$s(t)$	$\alpha[s(t)]$	$\rho[s(t)]$	$K^z[z_R, s(t)]$ (m^2 s^{-1})	$K^x[s(t)] = K^y[s(t)]$ (m^2 s^{-1})
A	0.05	6	45.0	250.0
B	0.1	6	15.0	100.0
C	0.2	4	6.0	30.0
D	0.3	4	2.0	10.0
E	0.4	2	0.4	3.0
F	0.5	2	0.2	1.0

For K_z, the classic formula used by Shir and Shieh (1974) was modified in the following way:

$$K_z[z,s(t)] = K_D[s(t)]\ z \exp\{-\rho[s(t)]\ z/H\} \tag{19}$$

Values of $\rho(s)$ are reported in Table 1 together with values of $K_z(z_R, s)$ (the vertical diffusion coefficient at level z_R), from which $K_D(s)$ is obtained as follows:

$$K_D(s) = z_R^{-1} K_z(z_R, s) \exp[\rho(s)\ z_R/H] \tag{20}$$

Table 1 also gives values for the horizontal dispersion coefficient $K_x(s) = K_y(s)$ which is assumed to be constant.

4.4 Simulation Results

The model described was used to simulate the concentration in the Venetian area in the period March–October 1973. The results were in general unsatisfactory and were particularly disappointing for cases when pollution episodes occurred. Two examples are reported in Figures 4 and 5. In both cases the discrepancies between the model and reality can be reasonably ascribed to input uncertainties, which range from the scarce knowledge about the wind field and the diffusive properties of the atmosphere to the very approximate and incomplete information about emissions.

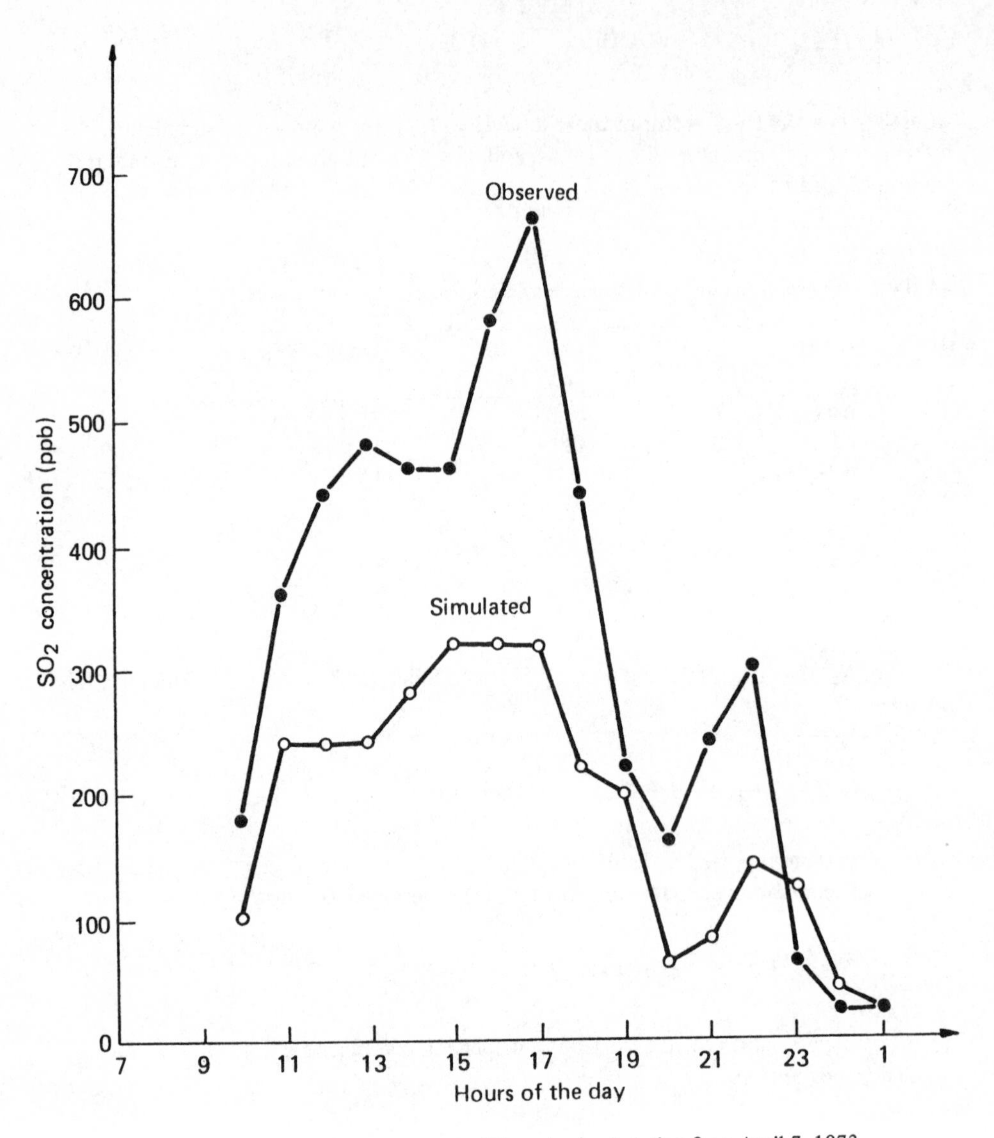

FIGURE 4 The simulation and measurement of the episode at station 9 on April 7, 1973.

5 THE NECESSITY OF A DIFFERENT APPROACH

It might be thought that an advection–diffusion model more sophisticated than the one described here could give a better account of real pollution phenomena. That this is not the case is shown by the application of Shir and Shieh's model to the Venice area (Marziano et al., 1979). Although the model is much more elaborate than the one described in this paper (it takes into account nonhomogeneous wind field, variable mixing height, surface roughness, etc.) the quality of the results presented by Marziano et al. is

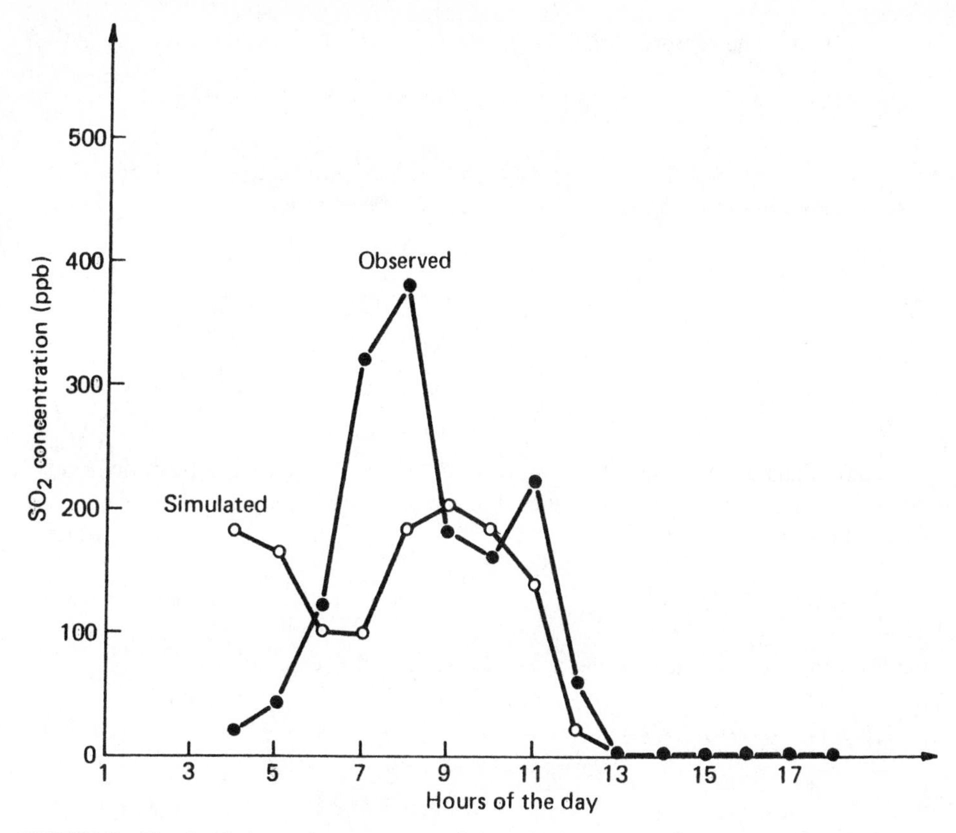

FIGURE 5 The simulation and measurement of the episode at station 30 on August 2, 1973.

comparable with that of the results given by the present model. Thus there is obviously a need for a different approach. As suggested by Bankoff and Hanzevack (1975), this different approach can be that of reformulating the model in a stochastic framework and applying the Kalman filtering technique. The first step in this procedure consists of interpreting the numerical scheme as a discrete dynamic system. This is quite straightforward owing to the nature of the algorithm used. In fact if we denote by the notation $|b^k_{ijm}|$ the row vector

$$|b^k(0,0,0), b^k(1,0,0), \ldots, b^k(I+1, J+1, M+1)|^T$$

and let

$$X(k) = |C^k_{ijm}| \; X^* = |C^*_{ijm}|, \text{ and } E(k) = |e^k_{ijm}|$$

where

$$e^k_{ijm} = \gamma_{ijm} Q_r \Delta t / \Delta x_r \Delta y_r \Delta z_r$$

then eqn. (13) takes the compact form

$$X^* = X(k) + E(k) \tag{21}$$

Similarly the concentration field X^{**} produced by the advection step can be obtained on the basis of eqns. (14) and (15) as

$$X^{**} = F_x(\alpha_{i,m}^k)X^* \tag{22}$$

where

$$\alpha_{i,m}^k = v_{x_m} \Delta t / \Delta x_i$$

is the Courant number. It is apparent from eqn. (22) that the matrix F is a function of the wind field and the atmospheric stability. For the diffusion step, eqn. (16) similarly produces a linear relationship

$$X^{****} = B_x [s(k)] X^{***} \tag{23}$$

so that if the intermediate fields are eliminated we obtain the relationship

$$X(k + 1) = \phi[v(k), s(k)] X(k) + \phi[v(k), s(k)] E(k) \tag{24}$$

where

$$\phi[v(k), s(k)] = F_x(\alpha_{i,m}^k) F_y(\alpha_{j,m}^k) B_x [s(k)] B_y [s(k)] B_z [s(k)]$$

It has been shown (Fronza et al., 1979) that the embedding of the discrete system (24) in a stochastic environment and the application of the Kalman filtering algorithm on the basis of the measurements supplied by a monitoring network produces a real-time predictor whose forecasting performance is a great improvement on that of the advection–diffusion model by itself.

REFERENCES

Bankoff, S.G. and Hanzevack, E.L. (1975). The adaptive filtering transport model for prediction and control of pollutant concentration in an urban airshed. Atmos. Environ., 9:793–808.

Christensen, O. and Prahm, L.P. (1976). A pseudospectral model for dispersion of atmospheric pollutants. J. Appl. Meteorol., 15:1284–1294.

Crank, J. and Nicolson, P. (1947). A practical method for numerical solution of partial differential equations of heat conduction type. Proc. Cambridge Philos. Soc., 43:50–67.

Desalu, A.A., Gould, L.A., and Schweppe, F.C. (1974). Dynamic estimation of air pollution. IEEE Trans. Autom. Control, 19:904–910.

Egan, B.A. and Mahoney, J.R. (1972). Numerical modeling of advection and diffusion of urban area source pollutants. J. Appl. Meteorol., 11:312–322.

Fronza, G., Spirito, A., and Tonielli, A. (1979). Real-time Prediction of SO_2 Pollution in Venice Lagoon Area. Part II: Kalman Predictor. RR-79-11. International Institute for Applied Systems Analysis, Laxenburg, Austria.

Lange, R. (1973). ADPIC – A three-dimensional computer code for the study of pollutant dispersal and deposition under complex conditions. Rep. UCRL-51462. Lawrence Livermore Laboratory, Livermore, California.

Marziano, G.L., Shir, C.C., Shieh, L.J., Sutera, A., Gianolio, L., and Ciprian, M. (1979). Study of the SO_2 distribution in Venice by means of an air quality simulation model. Atmos. Environ., 13:477–487.

Melli, P. (1976). An application of the Galerkin method to the eulerian–lagrangian treatment of time-dependent advection and diffusion of air pollutants. In Proc. Int. Conf. on Finite Elements for Water Resources, Princeton, New Jersey. Pentech Press, Plymouth, England.

Pasquill, F. (1974). Atmospheric Diffusion, 2nd edn., Horwood, Chichester.

Randerson, D. (1970). A numerical experiment in simulating the transport of sulfur dioxide through the atmosphere. Atmos. Environ., 4:615–632.

Richtmyer, R.D. and Morton, K.W. (1967). Difference Methods for Initial-value Problems. Interscience, New York.

Roberts, K.V. and Weiss, N.O. (1966). Convective difference schemes. Math. Comput., 20:272–299.

Runca, E. (1976). An efficient air quality *K* model. In Proc. NATO–CCMS Int. Tech. Meet. on Air Pollution Modeling and its Application, 7th, Airlie, Virginia.

Runca, E. and Sardei, F. (1975). Numerical treatment of time-dependent advection and diffusion of air pollutants. Atmos. Environ., 9:69–80.

Sardei, F. and Runca, E. (1976). An efficient numerical scheme for solving time-dependent problems of air pollution advection and diffusion. In Proc. IBM Semin. on Air Pollution Modeling. IBM Italy Tech. Rep., Rome.

Shir, C.C., and Shieh, L.J. (1974). A generalized urban air pollution model and its application to the study of SO_2 distributions in the St. Louis Metropolitan Area. J. Appl. Meteorol., 13:185–204.

Sklarew, R.C., Fabrick, A.J., and Prager, J.A. (1971). A particle-in-cell method for numerical solution of the atmospheric diffusion equation, and application to air pollution problems. Tech. Rep. 3SR-844. Systems, Science and Software, La Jolla, California.

Yanenko, N.N. (1971). The Method of Fractional Steps. Springer, Berlin.

Zannetti, P., Melli, P., and Runca, E. (1977). Meteorological factors affecting SO_2 pollution levels in Venice. Atmos. Environ., 11:605–616.

KALMAN PREDICTION OF SULFUR DIOXIDE EPISODES

G. Fronza and A. Spirito
Centro Teoria dei Sistemi, Milan (Italy)

A. Tonielli
Istituto Automatica, Università Bologna, Bologna (Italy)

1 INTRODUCTION

In general terms, real-time prediction of air pollution means forecasting of future ground-level concentrations on the basis of current information about meteorology, scheduled future emissions, and the concentrations themselves. Usually a mathematical real-time predictor (the Kalman predictor; see for instance Kalman, 1960 and Jazwinski, 1970) is derived from a stochastic dynamic model describing the dispersion of the pollutant in the airshed. Basically, such models belong to one of the following classes.

(a) "Black-box" models of the AutoRegressive Integrated Moving Average with exogenous inputs (ARIMAX) type, (see for instance Merz et al., 1972; McCollister and Wilson, 1975; Chock et al., 1975; Tiao et al., 1975; and Finzi et al., 1977a, b, 1978, 1979). In an ARIMAX representation, pollutant concentrations (or related variables) at a particular instant are expressed as a linear combination of previous concentration values plus a linear combination of present and previous emissions plus random terms (noise), which are specified statistically. Moreover, the coefficients of both linear combinations depend on meteorological variables.

(b) "Grey-box" models consisting of a stochastic version of a numerical solution scheme of the advection–diffusion equation (see Bankoff and Hanzevack, 1975; Desalu et al., 1974; and Sawaragi and Ikeda, 1974).

In the present paper we follow approach (b); more precisely, we describe the result of an application of the Kalman predictor (derived from a grey-box stochastic model) to the forecasting of SO_2 episodes in the Venetian lagoon area. A stochastic version of the Carlson–Crank–Nicolson scheme already shown by Runca et al. (1982) is illustrated in Section 2 together with the derived Kalman predictor.

In Section 3 we describe the forecasting performance in the Venetian case together with some ad hoc procedures required for actual implementation of the predictor.

2 THE KALMAN PREDICTOR

The starting point for the development of the recursive prediction formula is the numerical solution scheme of the pollutant advection–diffusion equation. Runca et al. (1982) have already shown that such a scheme can be written in the form of a discrete dynamic system:

$$X(k+1) = \phi\,[v(k),s(k)]\,X(k) + \phi\,[v(k),s(k)]\,E(k) \tag{1}$$

$$y(k) = HX(k) \tag{2}$$

where $X(k)$ is the vector of the kth hourly average concentrations at all the grid points of the solution scheme, $E(k)$ is the emission vector of the kth source, $y(k)$ is the vector of the measured concentrations in the kth hour, $\phi\,[v(k), s(k)]$ is a suitable matrix depending on the wind vector $v(k)$ and the stability class $s(k)$, and H is a suitable binary matrix.

The embedding of the discrete system (1)–(2) into a stochastic environment simply requires the addition of random terms in its two relationships to represent the "inaccuracies" of the model itself. Thus the stochastic version of the system (1)–(2) is given by

$$X(k+1) = \phi\,[v(k), s(k)]\,X(k) + \phi\,[v(k), s(k)]\,E(k) + n(k) \tag{1'}$$

$$y(k) = HX(k) + w(k) \tag{2'}$$

where $n(k)$ is a stochastic term ("process noise") which accounts for all the sources of disagreement between the model and the actual dynamics of the pollution phenomenon (i.e. for physical inputs neglected in the advection–diffusion equation such as rain or chemical reactions, for errors introduced by the assignment of parameter values, for errors due to the model structure, numerical inaccuracies, etc.) and $w(k)$ is a stochastic term ("measurement noise") which accounts for errors in the measurements.

The random processes $[n(k)]_k$ and $[w(k)]_k$ are commonly assumed to be zero-mean white noises, that is, they have a correlation structure of the type

$$E\,[n(k)n^{\mathrm{T}}(k+\tau)] = \begin{cases} Q(k) & \tau = 0 \\ 0 & \tau \neq 0 \end{cases}$$

$$E\,[w(k)w^{\mathrm{T}}(k+\tau)] = \begin{cases} R(k) & \tau = 0 \\ 0 & \tau \neq 0 \end{cases}$$

where $E(\,)$ is an expectation operator. The problem of evaluating $Q(k)$ and $R(k)$ is discussed below.

The stochastic model (1′)–(2′) can be used for real-time pollution forecasting, i.e. for predicting, at the beginning of each time interval, future concentration levels on the basis of current information about concentrations, emission, and meteorology. Specifically, the recursive one-step-ahead forecast algorithm (the Kalman predictor) derived from the model (1′)–(2′) is given (see for instance Jazwinski, 1970) by

$$\hat{X}(k|k) = \hat{X}(k|k-1) + G(k)[y(k) - H\hat{X}(k|k-1)] \tag{3}$$

$$\hat{X}(k+1|k) = \phi[v(k),s(k)]\,\hat{X}(k|k) + \phi[v(k),s(k)]\,E(k) \tag{4}$$

where $\hat{X}(k+1|k)$ is the prediction of $X(k+1)$ made at time k (i.e. at the beginning of the kth time interval), and $\hat{X}(k|k)$ is the filtered state, i.e. the a posteriori (at time k) estimate of $X(k)$ on the basis of the new available datum $y(k)$. This estimate is given by eqn. (3) as a correction of the previous forecast $\hat{X}(k|k-1)$ and is introduced in eqn. (4) instead of $\hat{X}(k|k-1)$ in order to give a better prediction of $X(k+1)$. $G(k)$ is the Kalman gain given by

$$G(k) = P(k|k-1)H^{\mathrm{T}}[HP(k|k-1)H^{\mathrm{T}} + R(k)]^{-1} \tag{5}$$

where $P(k|k-1)$ is the forecasting error covariance matrix:

$$P(k|k-1) = E\{[\hat{X}(k|k-1) - X(k)]\,[\hat{X}(k|k-1) - X(k)]^{\mathrm{T}}\}$$

In turn, this covariance matrix is recursively evaluated through the equations

$$P(k|k) = P(k|k-1)\{I - H^{\mathrm{T}}[HP(k|k-1)H^{\mathrm{T}} + R(k)]^{-1}HP(k|k-1)\} \tag{6}$$

$$P(k+1|k) = \phi[v(k),s(k)]\,P(k|k)\phi^{\mathrm{T}}[v(k),s(k)] + Q(k+1) \tag{7}$$

The r-step-ahead prediction ($r = 2,3,\ldots$) is obtained recursively from

$$\hat{X}(k+r|k) = \phi[v(k+r-1),s(k+r-1)]\,\hat{X}(k+r-1|k)$$

$$+\,\phi[v(k+r-1),s(k+r-1)]\,E(k+r-1) \tag{8}$$

The actual implementation of the Kalman predictor (3)–(8) raises a number of conceptual and practical problems which are now discussed in detail.

2.1 The Assignment of $Q(k)$ and $R(k)$

The a posteriori correction of the previous forecast $\hat{X}(k|k-1)$ to $\hat{X}(k|k)$ [a better initial state for the new prediction step (11)] is made in eqn. (3) by weighting the new datum $y(k)$ through the Kalman gain $G(k)$. In turn, this weight matrix depends, in view of eqns. (5)–(7), on the noise intensities $Q(k)$ and $R(k)$. Finally, at every forecasting step the Kalman predictor corrects ("filters") the initial state of the step by taking into account the noise intensities. Clearly, $Q(k)$ and $R(k)$ must be regarded as input data to the filter [eqns. (3) and (5)–(8)]; i.e., they must be evaluated before the filtering is performed. In principle $R(k)$ can be obtained from an analysis of the accuracy of the measurement system while $Q(k)$ can be obtained from considerations of fitting between the numerical scheme and the real world.

However, this is not done in practice. Instead $Q(k)$ and $R(k)$ are usually estimated by recursive algorithms based on an a posteriori analysis at each time step of the performance of the predictor at previous time steps. There are different types of adaptive Kalman predictors, namely Kalman predictors supplied with a recursive algorithm for the a posteriori estimation of $Q(k)$ and $R(k)$. Unfortunately the most rigorous predictors (see for instance Mehra, 1970) cannot be applied in the present case because of the nonstationarity of the system (1′)–(2′) (more precisely, because the matrix $\phi[v(k),s(k)]$ is not the same for every k).

Thus in the application described in Section 3 we used a heuristic adaptive approach which represents a slight generalization of a procedure used by Jazwinski (1969).

2.2 The Treatment of Emission Uncertainties

The adaptive mechanisms mentioned earlier (the updating of the noise intensities at each time step) are usually too weak to allow for accurate forecasting of pollution episodes when the episodes are due to conspicuous (but unknown to the predictor) emission enhancements. Usually a more robust correction is obtained by introducing an additive term $p(k)$ in eqn. (1′):

$$X(k+1) = \phi[v(k),s(k)]\,X(k) + \phi[v(k),s(k)]\,E(k) + p(k) + n(k) \tag{1''a}$$

with the dynamics

$$p(k+1) = p(k) \tag{1''b}$$

This procedure is called state enlargement and the Kalman predictor derived from the model (1″a)–(1″b)–(2′) is called an extended predictor (see for instance Jazwinski, 1970). Although they are often effective, extended (adaptive) Kalman predictors require the dimension of the state of the stochastic system to be doubled [from eqn. (1″a), $p(k)$ has the same dimension as $X(k)$] and correspondingly the computational burden increases heavily.

Hence a simpler and less expensive, though more approximate and heuristic, recursive adjustment of emission inputs was considered in the present work. More precisely, eqn. (1′) was modified as follows:

$$X(k+1) = \phi[v(k),s(k)]\,X(k) + \phi[v(k),s(k)]\,\theta(k)E(k) + n(k) \tag{1'''}$$

The scalar $\theta(k)$ is defined as

$$\theta(k) = \sum_l \hat{x}_l(k\,|\,k) \Big/ \sum_l \hat{x}_l(k\,|\,k-1) \tag{9}$$

where $\hat{x}_l(k\,|\,k)$ and $\hat{x}_l(k\,|\,k-1)$ are the l components of $\hat{X}(k\,|\,k)$ and $\hat{X}(k\,|\,k-1)$, respectively, and the summations are performed over all the components.

From eqn. (9), $\theta(k)$ is the ratio between the total "mass" of pollutant estimated a posteriori [i.e., after the arrival of the new measurement vector $y(k)$] and the mass previously forecast. If the two masses do not coincide then an emission variation, occurring between time $k - 1$ and time k and not revealed to the predictor, is assumed and the emission for the subsequent step [the prediction of $X(k + 1)$ made at time k] is correspondingly modified.

2.3 Considerations of Computational Effort

From eqns. (3)–(8) the implementation of the Kalman predictor implies at each iteration step the manipulation of square matrices having an order equal to the number of grid points. Though the numerical scheme described by Runca et al. (1982) admits non-uniform grid spacing, the order of these matrices may easily be 1000, corresponding to an intolerable computational burden. In order to reduce this burden, the following procedure (due to Bankoff and Hanzevack) was used.

(a) A certain number of disconnected subregions of interest (i.e. subregions where episodes usually occur) was selected. Denote these subregions by $\mathcal{R}_1, \mathcal{R}_2, \ldots, \mathcal{R}_D$ and let $X_d(k)$ $(d = 1, 2, \ldots, D)$ be the vector of components of $X(k)$ corresponding to the grid points of $\mathcal{R}_d$ [$X_d(k)$ may, for example, be a subvector of 30 components]. For simplicity, assume for a moment that the order of the components of $X(k)$ is rearranged, so that $X(k)$ can be partitioned as

$$X(k) = |X_1^T(k)\ X_2^T(k)\ \ldots\ X_D^T(k)\ X_{out}^T(k)|^T$$

where $X_{out}(k)$ is the vector of the components of $X(k)$ corresponding to grid points outside all the subregions.

(b) Apply the filter [eqns. (3) and (5)–(7)] only to each subregion; i.e. evaluate filtered subvectors $\hat{X}_1(k|k), \hat{X}_2(k|k), \ldots, \hat{X}_D(k|k)$. Each of these D applications of the filtering procedure involves the manipulation of matrices of reasonable dimensions.

(c) Modify the forecasting step (4) [and the corresponding step (8)] by

$$\hat{X}(k + 1|k) = \phi[v(k), s(k)]\,\tilde{X}(k|k) + \phi[x(k), s(k)]\,E(k) \tag{4'}$$

where

$$\tilde{X}(k|k) = \ \hat{X}_1^T(k|k)\ \hat{X}_2^T(k|k)\ \ldots\ \hat{X}_D^T(k|k)\ \hat{X}_{out}^T(k|k-1)|^T \tag{10}$$

From eqns. (4′) and (10) it is found that the procedure (a)–(c) simply corresponds to the filtering of only one state subvector at each step of the Kalman predictor.

There is clearly a danger inherent in the method. As is apparent from eqn. (10), the initial state $\tilde{X}(k|k)$ of the forecasting step (4′) may turn out to be a very "irregular" concentration field since some state components have been filtered and some have not. In particular, strong variations may result between the components corresponding to subregion boundaries and components corresponding to grid points immediately outside. Hence "artificial" high gradients may be introduced, with negative effects on the forecasting step (4′) [which is merely a step in the numerical scheme, described by Runca et al. (1982), with initial field $\tilde{X}(k|k)$]. Whether the distortion caused by partial filtering is relevant or not can be ascertained only by simulation of the Kalman predictor on the real case (see Section 3).

2.4 Meteorological Input to the Predictor

From eqns. (4) and (8), the forecast of r-step-ahead concentration fields made at the beginning of the kth time interval requires knowledge of the emission, the wind field, and the stability inputs for the time intervals $[k\Delta t,(k+1)\Delta t]$, $[(k+1)\Delta t,(k+2)\Delta t],\ldots,$ $[(k+r-1)\Delta t,(k+r)\Delta t]$.

The required information about emission in these future periods should in principle be available since emission is a decision variable of polluters, who are assumed to collaborate or to be forced to collaborate with the prediction (in any case, if incorrect information is supplied, the forecast quality is safeguarded by the correction mechanism described in Section 2.2).

However, the future meteorology is not obviously known. There are basically two possible approaches for supplying the wind and stability inputs required by the concentration predictor: (i) set up a meteorological predictor whose forecasts are introduced as inputs into eqns. (4) and (8); (ii) simply postulate a persistent meteorology (the wind and stability in the future will have the same values as at present).

Clearly the forecasting performance of the concentration predictor under approach (ii) is a lower bound since it corresponds to the most approximate treatment of the meteorological inputs. In contrast, the upper bound of the forecasting performance corresponds to a situation where the wind and stability inputs are supplied to eqns. (4) and (8) by a perfect meteorological predictor (the forecast wind and stability are always equal to their future true values).

The performance under approach (i) is expected to lie within the two bounds, its distance from the upper bound obviously depending on the quality of the meteorological forecast.

This analysis of the sensitivity of concentration prediction to the treatment of meteorological inputs was actually performed in the Venetian Lagoon study, as illustrated in detail in the next section.

3 THE APPLICATION OF THE PREDICTOR TO THE VENETIAN LAGOON STUDY

The air-pollution problem and monitoring network for the Venetian Lagoon have already been described by Runca et al. (1982) together with the three-dimensional model used in this application.

A stochastic embedding procedure, quite similar to that illustrated in Section 2, holds for the numerical solution scheme of the three-dimensional advection–diffusion equation. The resulting stochastic model is rewritten here by making it evident that the wind speed has two (horizontal) components in this case:

$$X(k+1) = \phi[v(k),d(k),s(k)]\,X(k) + \phi[v(k),d(k),s(k)]\,E(k) + n(k) \tag{11}$$

$$y(k) = HX(k) + w(k) \tag{12}$$

where $v(k)$ and $d(k)$ respectively denote the vector of wind intensities at different levels and the vector of wind direction. For the Kalman predictor derived from the system (11)–(12) we must make specifications with regard to the points raised in Sections 2.3 and 2.4.

The locations of the three grid subregions mentioned in Section 2.3 (i.e. subregions where the concentration field is filtered) are illustrated in Figure 1. The shaded areas of Figure 1 represent the bases of the three subregions at ground level ($m = 1$) and correspond to the most polluted zones. Along the vertical axis each subregion reaches the level $m = 3$, i.e., has an extension of two layers above the ground. The three subregions correspond to 18, 18, and 16 state variables, respectively.

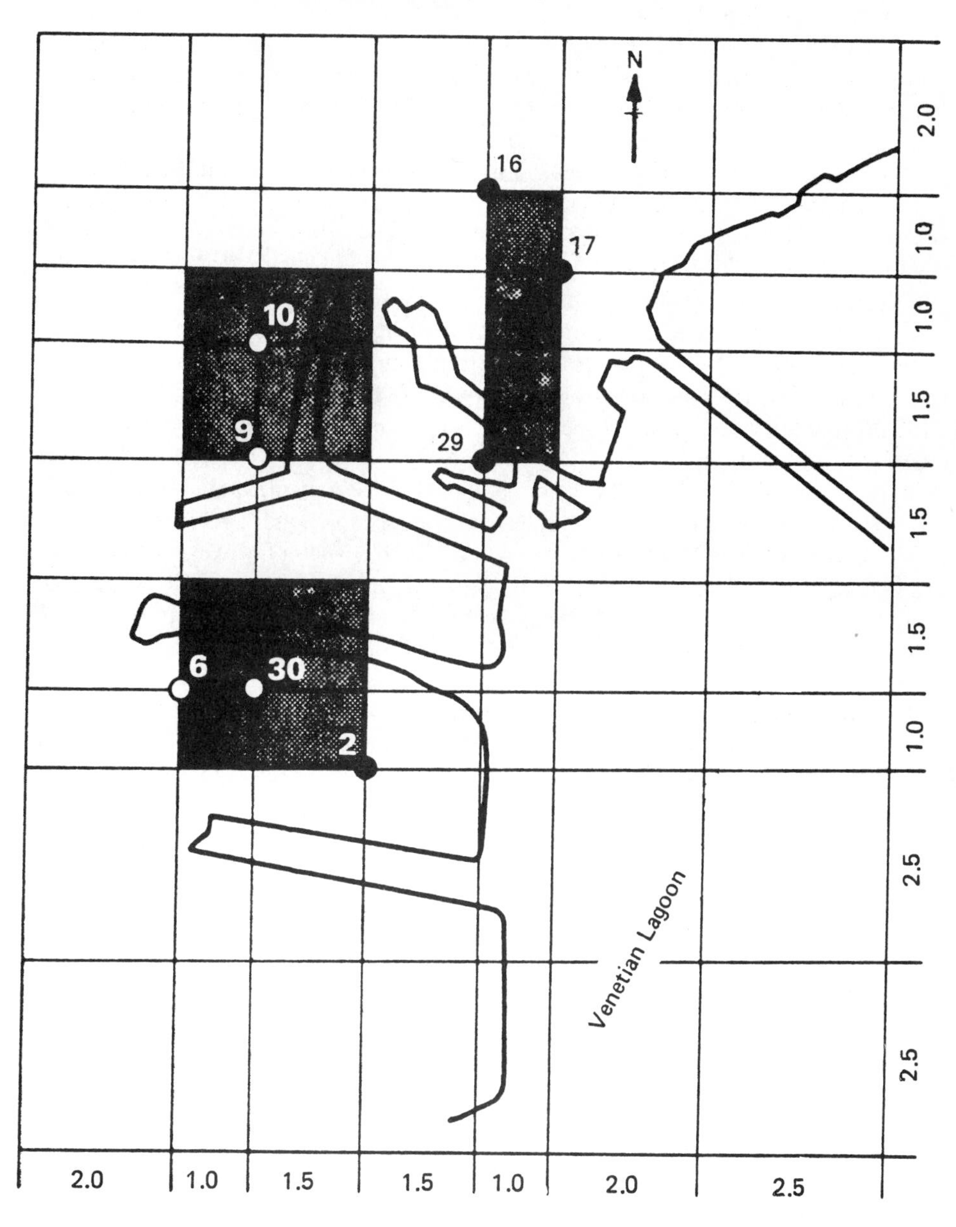

FIGURE 1 The ground plan of the discretization grid (distances in kilometers) showing stations (•) and subregions (shaded areas).

With regard to Section 2.4, the Kalman predictor of future concentrations was run in correspondence with three different types of meteorological input.

(i) Inputs given by a mathematical meteorological predictor. More precisely, for $f = 1, 2, \ldots, r-1$ a forecast $\hat{s}(k+f\,|\,k)$ (made at time k for the stability class at time $k+f$) was obtained in accordance with the simple probabilistic criterion

$$\hat{s}(k+f\,|\,k) \to \max_{s(k+f)} \mathrm{Prob}\,[s(k+f)\,|\,s(k-1), q]$$

where $\mathrm{Prob}\,[s(k+f)\,|\,s(k-1), q]$ is the probability of having class $s(k+f)$ at time $k+f$, given the information that the class has been $s(k-1)$ in the interval $[(k-1)\Delta t, k\Delta t]$ and that time $k\Delta t$ is the qth hour of the day ($q = 1, 2, \ldots, 24$).

Similarly the wind direction sector was forecast in accordance with the criterion

$$\hat{d}(k+f\,|\,k) \to \max_{d(k+f)} \mathrm{Prob}\,[d(k+f)\,|\,d(k-1), q]$$

Finally the wind intensity at the meteorological station was forecast by means of an autoregressive moving average predictor (see Box and Jenkins, 1970). From this forecast, predictions $\hat{v}(k+f\,|\,k)$ of the whole future profiles of wind intensities were obtained through the power law illustrated by Runca et al. (1982).

(ii) Inputs given by assuming persistent meteorology $[v(k+f) = v(k-1), d(k+f) = d(k-1), s(k+f) = s(k-1)]$.

(iii) Inputs given by assuming a perfect meteorological predictor, i.e. true inputs.

The 4-h-ahead forecasting performance under the three conditions of input treatment is shown in Figure 2 for the episode of April 7, 1973, and in Figure 3 for the episode of August 2, 1973. As expected, approach (i) gives a performance intermediate between those for approaches (ii) and (iii) but very near to the ideal situation for approach (iii). In fact the correlations between the forecast and the true concentration data were 0.90, 0.32, and 0.92 for the cases in Figures 2(a), 2(b), and 2(c), respectively, and 0.76, 0.50, and 0.77 for the cases in Figures 3(a), 3(b), and 3(c), respectively. A comparison of Figures 2 and 3 with the corresponding performances of the numerical scheme alone (Figures 4 and 5, respectively in Runca et al. (1982)) shows clearly the improvement of quality.

For the remainder of the concentration field it must first be recalled that the episodes mentioned here are characterized by a strongly nonuniform field – to be precise, by a relevant peak at one station (station 9 for the episode of April 7, 1973, and station 30 for the episode of August 2, 1973) and by relatively low concentrations elsewhere. These low concentrations were satisfactorily forecast by the Kalman predictor, i.e. the "artificial gradient effect" caused by subregion filtering (see Section 2.3) did not occur.

However, there was a certain overestimation of the field outside the station affected by the pollution peak. This was due to the correction mechanism for the emissions (see Section 2.2) which is based on the multiplicative scalar coefficient $\theta(k)$ and hence results in the simultaneous enhancement of all the emissions. Thus the existence of an episode around a particular station causes the correction mechanism to increase all the emissions and consequently to increase the whole forecast field. Of course this effect could be avoided by setting up a suitable selective mechanism for correction of the emissions.

In nonepisode situations, the predictor also performed well, but this is not a particularly significant result. Finally, with regard to computation times, each 4-h-ahead forecast required approximately 3 min on an IBM 370 computer.

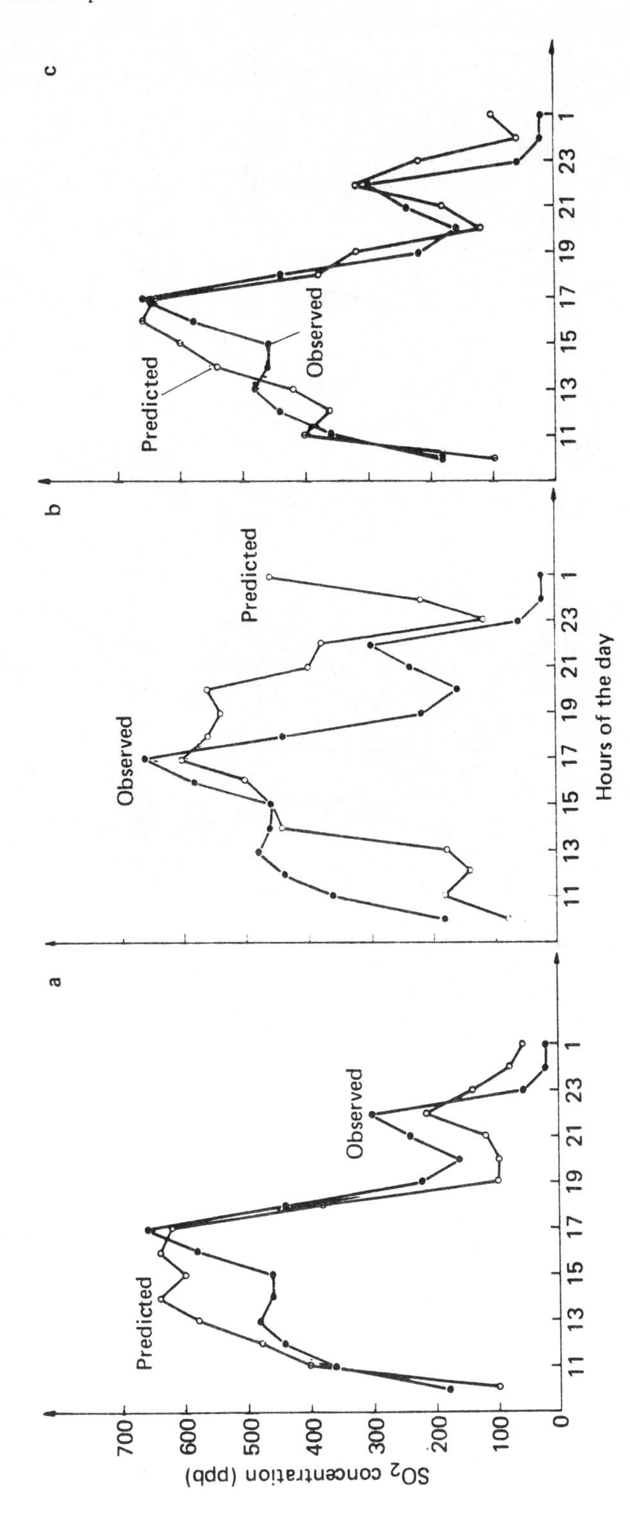

FIGURE 2 4-h-ahead Kalman episode prediction with (a) forecast, (b) persistent, and (c) true meteorological inputs (April 7, 1973).

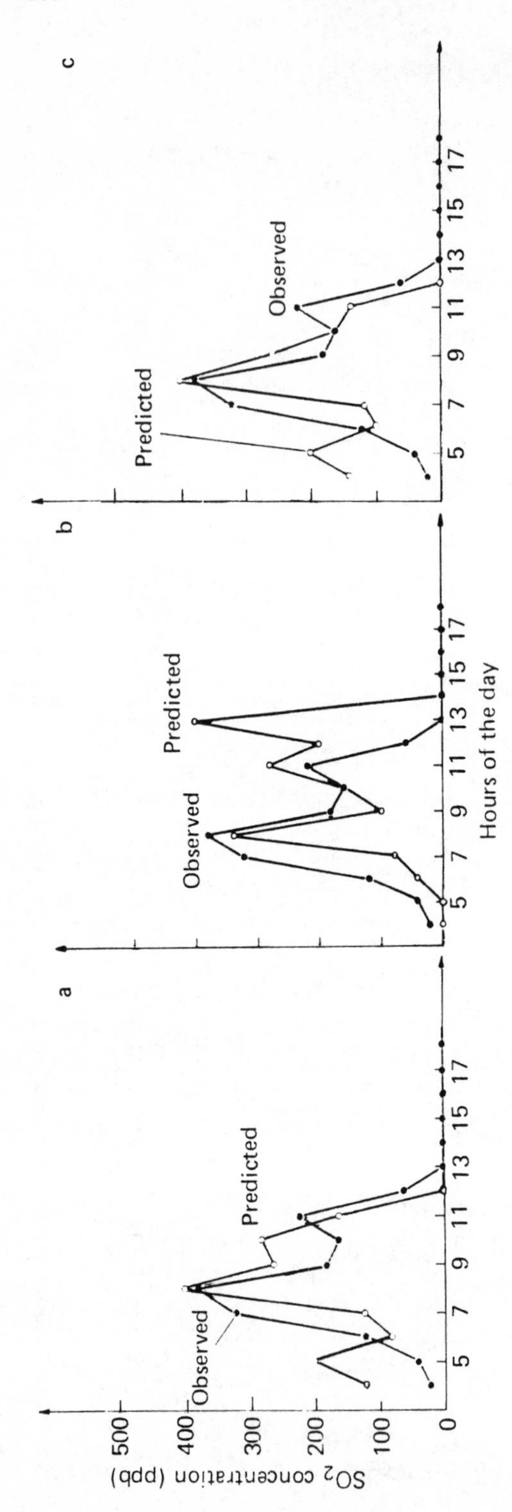

FIGURE 3 4-h-ahead Kalman episode prediction with (a) forecast, (b) persistent, and (c) true meteorological inputs (August 2, 1973).

REFERENCES

Bankoff, S.G. and Hanzevack, E.L. (1975). The adaptive filtering transport model for prediction and control of pollutant concentration in an urban airshed. Atmos. Environ., 9:793–808.

Box, G.E.P. and Jenkins, G.M. (1970). Time Series Analysis, Forecasting and Control. Holden–Day, San Francisco, California.

Chock, E.P., Terrel T.R., and Levitt, S.B. (1975). Time series analysis of Riverside, California, air quality data. Atmos. Environ., 9:978–989.

Desalu, A.A., Gould, L.A., and Schweppe, F.C. (1974). Dynamic estimation of air pollution. IEEE Trans. Autom. Control, 19:904–910.

Finzi, G., Fronza, G., and Rinaldi, S. (1978). Stochastic modeling and forecast of the dosage area product. Atmos. Environ., 12:831–838.

Finzi, G., Fronza, G., Rinaldi, S., and Zannetti, P. (1977a). Modeling and forecast of the dosage population product in Venice. In Proc. IFAC Symp. on Environmental Systems, Design, and Control, Kyoto, Japan.

Finzi, G., Fronza, G., and Spirito, A. (1977). Univariate stochastic models and real-time predictors of daily SO_2 pollution in Milan. In Proc. NATO–CCMS Tech. Meet. on Air Pollution Modeling, 7th, Louvain-la-Neuve, Belgium.

Finzi, G., Zannetti, P., Fronza, G., and Rinaldi, S. (1979). Real-time prediction of SO_2 concentration in the Venetian lagoon area. Atmos. Environ., 13:1249–1255.

Jazwinski, A.H. (1969). Adaptive filtering. Automatica, 5:475–495.

Jazwinski, A.H. (1970). Stochastic Processes and Filtering Theory. Academic Press, New York.

Kalman, R.E. (1960). A new approach to linear filtering and prediction theory. Trans. ASME, Ser. D, 82:17–25,

McCollister, G.M. and Wilson, J.R. (1975). Linear stochastic models for forecasting daily maxima and hourly concentrations of air pollutants. Atmos. Environ., 9:417–423.

Mehra, R.K. (1970). On the identification of variances and adaptive Kalman filtering. IEEE Trans. Autom. Control, 15:175–184.

Merz, P.H., Painter, L.J., and Ryason, P.R. (1972). Aerometric data analysis – time series analysis and forecast of an atmospheric smog diagram. Atmos. Environ., 6:319–322.

Runca, E., Melli, P., and Spirito, A. (1982). A *K* model for simulating the dispersion of sulfur dioxide in an airshed. In G. Fronza and P. Melli (Editors), Mathematical Models for Planning and Controlling Air Quality. Pergamon Press, Oxford. (This volume, p. 147.)

Sawaragi, Y. and Ikeda, S. (Editors) (1974). Proc. Symp. on Modeling of the Control and Prediction of Air Pollution, Kyoto, Japan.

Tiao, G.C., Box, G.E.P., and Hamming, W.J. (1975). A statistical analysis of the Los Angeles ambient carbon monoxide data. J. Air Pollut. Control Assoc., 25:1129–1136.

THE COST OF A REAL-TIME CONTROL SCHEME FOR SULFUR DIOXIDE EMISSIONS

P. Melli
Centro Scientifico IBM, Rome (Italy)

P. Bolzern, G. Fronza, and A. Spirito
Centro Teoria dei Sistemi, Milan (Italy)

1 INTRODUCTION

A large number of mathematical models are now available to air-quality decision makers. Almost all the models concern long-term management (i.e. planning) problems such as the allocation and design of polluting sources (see for instance Emanuel et al., 1978; and Guldmann and Shefer, 1976), the choice of pollutant treatment levels, i.e. the levels of permanent reductions of emission rates (see for instance Atkinson and Lewis, 1974; Seinfeld and Kyan, 1971), and the definition of taxation or other regulation schemes.

However, planning is just one side of an air-quality management problem and in some situations it may not even be the most relevant aspect, for the following reasons.

(a) Many cases of air pollution are due to already-existing sources; i.e., there is no source allocation and design problem.

(b) Apart from technological difficulties, a pollution regulation scheme which is based only on the permanent abatement of emission by treatment plants may meet with relevant drawbacks with regard to either costs or effectiveness. In fact, if emission reductions are established so that the most severe pollution episodes are flattened, the cost of treatment is likely to be extremely high. However, if the treatment is moderate, average pollution will be lowered but severe meteorological situations will still be likely to bring concentrations well beyond admissible levels.

An approach which can considerably reduce the drawbacks mentioned in (b) consists of replacing permanent smoke treatment by (or, at least, combining it with) real-time emission control (see for instance Shepard, 1970; and Leavitt et al., 1971). In general, real-time control is a sequential short-term management procedure which at the beginning of each time step (each hour, say) is based on the following operations.

(i) Using a monitoring network, collect the present and recent past values of pollutant concentrations and significant meteorological variables in the area under consideration.

(ii) On the basis of this information about the meteorological situation [and, possibly, of synoptic forecasts (Barbieri et al., 1979)] predict future values (for, say, the next 2 or 4 h) of the meteorological variables. This can be accomplished either by running a mathematical predictor (Bonivento and Tonielli, 1982) on a computer or, more simply, by experience.

(iii) Predict future concentration levels on the basis of the information about concentrations described in point (i), the forecast meteorology discussed in point (ii), and scheduled future emission rates. Again, this can be accomplished either by running a mathematical concentration predictor (see for instance Finzi et al., 1978, 1979; Bankoff and Hanzevack, 1975) or, more simply, by experience.

(iv) If the predicted concentrations exceed preassigned levels, reduce the scheduled emissions in accordance with some control policy.

As is clear from points (i)–(iv) above, the philosophy of real-time control is to take action only in a situation of a forthcoming "episode" (actually, of a forecast forthcoming episode). The control action is therefore discontinuous over time, thus yielding a conspicuous cost saving with respect to permanent treatment.

Naturally, a scheme like that outlined in points (i)–(iv) can be set up practically only where a limited number of sources have to be controlled (in an industrial area), although there have already been interesting applications to cases of urban pollution in Japan. Moreover, a prerequisite of the scheme is the establishment of reference pollution levels [the "preassigned levels" mentioned in point (iv)] for the control action.

In the present paper we describe a case study of the real-time control of SO_2 pollution from the industrial area of Porto Marghera in the Venetian Lagoon region. The details [the monitoring network for step (i), the meteorological predictor for step (ii), the concentration predictor for step (iii), and the reduction policy for step (iv)] are specified in the next section. In particular, the control action is assumed to be emission abatement by fuel substitution, under the constraint of maintaining the scheduled production for each polluting plant. The results of the analysis are summarized in Section 3 in terms of cost-effectiveness curves. Specifically, the reference concentration level mentioned in (iv) is used as a parameter. For each value of this parameter the cost of real-time control (measured by the percentage of extra expenditure caused by the use of low-sulfur fuel) is reported versus the effectiveness of the control policy (measured by a properly defined index).

2 DESCRIPTION OF THE REAL-TIME CONTROL SCHEME

The characteristics of the SO_2 pollution problem in the Venetian Lagoon area have already been illustrated by Runca et al. (1982). The details of the present application of the general real-time control scheme (i)–(iv) (see Section 1) are as follows.

2.1 The Monitoring Network

The network required for the real-time collection of the SO_2 concentrations and the relevant meteorological data has also been described by Runca et al. (1982).

2.2 The Meteorological Predictor

The predictors of the wind and the stability class required for the implementation of the concentration predictor (Section 2.3) have already been illustrated by Fronza et al. (1982), while Bonivento and Tonielli (1982) have given details of the autoregressive moving average wind predictor.

2.3 The Concentration Predictor

The concentration predictor derived from the "stochastic version" of a Carlson–Crank–Nicolson scheme for the advection–diffusion equation has also been described by Fronza et al. (1982).

2.4 The Real-Time Control Policy

Let i ($i = 1,2,\ldots,N$) denote the source index and consider the instant $k\Delta t$ ($k = 0,1,\ldots;\ \Delta t = 1$ h), i.e., the beginning of the $(k+1)$th hour.

First it is assumed that at each source the actual performance of the control action takes 1 h. More precisely, an operation of fuel replacement decided at $k\Delta t$ is effective from $(k+1)\Delta t$; i.e., it displays its effect in emission reduction only in the $(k+2)$th hour. Because of this assumption (which is commented on later) a control action decided at time $k\Delta t$ consists of a reduction of $Q^i(k+2)$ ($i = 1,2,\ldots,N$), (the SO_2 emission scheduled by the ith source for the $(k+2)$th hour) and not of a reduction of $Q^i(k+1)$.

The control policy is specified as follows. For $r = 2,3,4$,

$$\hat{x}_{\mathrm{M}}(k+r\,|\,k)$$

is the maximum ground-level hourly concentration forecast at time $k\Delta t$ by the concentration predictor for the $(k+r)$th interval and

$$\Delta_{\mathrm{M}}(k+r\,|\,k) = \begin{cases} [\hat{x}_{\mathrm{M}}(k+r\,|\,k) - c]/\hat{x}_{\mathrm{M}}(k+r\,|\,k) & \text{if positive} \\ 0 & \text{otherwise} \end{cases} \qquad (1)$$

where c is the preassigned reference concentration value.

The following control action is then assumed: reduce the emission $Q^i(k+2)$, scheduled for the $(k+2)$th hour by source i, by a percentage $\alpha(k+2)$ given by

$$\alpha(k+2) = \sum_{r=2}^{4} p_r \Delta_{\mathrm{M}}(k+r\,|\,k) \tag{2}$$

In eqn. (2), p_r is a preassigned weight such that $0 \leqslant p_r \leqslant 1$ and $\Sigma_{r=2}^{4}\, p_r = 1$. Therefore, since $0 \leqslant \Delta_{\mathrm{M}}(k+r\,|\,k) < 1$ in view of eqn. (1), from eqn. (2) it is always the case that $0 \leqslant \alpha(k+2) < 1$ (actually, the range of $\alpha(k+2)$ is narrower because of the production constraint (see Section 2.4.1)).

Control policy (2) is explained and completed by the following comments.

(a) From eqn. (2), $\alpha(k+2)$ is positive whenever (at least) one of the three $\Delta_{\mathrm{M}}(k+r\,|\,k)$ is greater than zero; i.e., [see eqn. (1)] a reduction of $Q^i(k+2)$ is made (via fuel replacement starting at time $k\Delta t$ because of the delay due to operation times) whenever (at least) one forecast of maximum concentration made at time $k\Delta t$ exceeds the reference level c.

(b) The three forecasts $\hat{x}_{\mathrm{M}}(k+2\,|\,k), \hat{x}_{\mathrm{M}}(k+3\,|\,k), \hat{x}_{\mathrm{M}}(k+4\,|\,k)$ clearly exhibit decreasing reliabilities. They must therefore affect the decision $\alpha(k+2)$ through decreasing weights; i.e., $p_2 > p_3 > p_4$ must hold.

(c) The control action $\alpha(k+2)$ does not depend on the source index i; i.e., the percent reduction is the same for all sources. This uniformity may appear to be too great a simplification since the detail of the information supplied by the concentration predictor (Fronza et al., 1982) seems to allow selective control of the sources. However, in the Venetian case discussed in Section 3 the reliability of the data on the scheduled emissions did not warrant the application of more sophisticated control policies.

(d) The forecast horizon for the control action (2) has been taken as 4 h whereas the delay due to operation (in other words, fuel replacement) times has been assumed to be 1 h. However, by proper modification of indexes in eqns. (1) and (2) the analysis can be adapted to longer forecast horizons as well as to longer (or shorter) operation times. Of course, the prediction reliability may become poor if the forecast horizon is too large.

(e) In view of eqns. (1) and (2), the decision is based on forecast maximum ground-level concentrations. However, other variables (spatial averages or similar) can be taken into account as guidelines for deciding the control action.

2.4.1 Actual Implementation of the Control Action and Extra-Cost Evaluation

In this section the operation of fuel replacement necessary to obtain the emission reduction $\alpha(k+2)$ is specified in terms of the required fuel quantities. In the following discussion the indexes HS (High Sulfur) and LS (Low Sulfur) will denote the standard fuel used by all the polluters and the cleaner fuel required for the control action, respectively.

First, if t_{HS} and t_{LS} denote the sulfur contents of the two fuels, the quantities (in tons) $q^i_{\mathrm{HS}}(k+2)$ and $q^i_{\mathrm{LS}}(k+2)$ to be burnt in the $(k+2)$th hour must be such that

$$t_{\mathrm{HS}}q^i_{\mathrm{HS}}(k+2) + t_{\mathrm{LS}}q^i_{\mathrm{LS}}(k+2) = [1-\alpha(k+2)]\,Q^i(k+2) \tag{3}$$

because of the definition of $\alpha(k+2)$.

Moreover, it should be noted that $Q^i(k+2)/t_{\mathrm{HS}}$ is the quantity of fuel scheduled by the ith source for the $(k+2)$th hour. Since no change in the production of the ith plant is desired, this overall quantity must not be modified by the control action; i.e., it must be given by

$$q^i_{HS}(k+2) + q^i_{LS}(k+2) = Q^i(k+2)/t_{HS} \tag{4}$$

From eqns. (3) and (4) it is straightforward to obtain

$$q^i_{HS}(k+2) = [1 - \alpha(k+2) - t_{LS}/t_{HS}]\, Q^i(k+2)/(t_{HS} - t_{LS}) \tag{5'}$$

$$q^i_{LS}(k+2) = \alpha(k+2) Q^i(k+2)/(t_{HS} - t_{LS}) \tag{5''}$$

Note that because of the non-negativity of $q^i_{HS}(k+2)$ in eqn. (5′) the control action is constrained to be less than or equal to $1 - t_{LS}/t_{HS}$. In particular, $\alpha(k+2) = 1 - t_{LS}/t_{HS}$ corresponds to complete replacement of the HS fuel $[q^i_{HS}(k+2) = 0, q^i_{LS}(k+2) = Q^i(k+2)/t_{HS}]$, i.e. to the "strongest" possible substitution.

If c_{HS} and c_{LS} denote the unit costs of the two fuels, the overall fuel cost in the $(k+2)$th hour for the N plants under control action is

$$C^c(k+2) = \sum_{i=1}^{N} [c_{HS} q^i_{HS}(k+2) + c_{LS} q^i_{LS}(k+2)]$$

i.e., in view of eqns. (5′) and (5″)

$$C^c(k+2) = \{c_{HS}[1 - \alpha(k+2) - t_{LS}/t_{HS}] + c_{LS}\alpha(k+2)\} Q(k+2)/(t_{HS} - t_{LS}) \tag{6}$$

where $Q(k+2) = \Sigma_{i=1}^{N}\, Q^i(k+2)$ is the overall scheduled emission.

The fuel cost without control action would simply be

$$C^{nc}(k+2) = c_{HS} Q(k+2)/t_{HS} \tag{7}$$

The percent extra cost due to control action is

$$\gamma(k+2) = [C^c(k+2) - C^{nc}(k+2)]/C^{nc}(k+2) \tag{8}$$

From eqns. (6) and (7) and some simple but cumbersome computations we obtain

$$\gamma(k+2) = \alpha(k+2)(c_{LS}/c_{HS} - 1)/(1 - t_{LS}/t_{HS}) \tag{9}$$

This formula illustrates the dependence of the percent extra cost on both the intensity of the emission reduction and the LS–HS unit-cost and sulfur-content ratios.

3 RESULTS IN THE VENETIAN CASE

The sequential real-time emission-control scheme described in the previous section was applied to episode smoothing in the Venetian Lagoon area.

As is clear from eqns. (1) and (2), the dependence of the control action on the reference concentration c is relevant. Hence the analysis was carried out using c as a parameter. Furthermore [see eqn. (2) again] two different triplets of p weights were considered: $p' = \{1;0;0\}$ and $p'' = \{0.5;0.3;0.2\}$. Finally two types of LS fuels (denoted by the indexes LS1 and LS2) were tested separately. The fuels correspond to the LS–HS sulfur-content ratios

$$t_{LS1}/t_{HS} = 0.33 \quad \text{and} \quad t_{LS2}/t_{HS} = 0.10 \tag{10}$$

and to the LS–HS unit-cost ratios

$$c_{LS1}/c_{HS} = 1.1 \quad \text{and} \quad c_{LS2}/c_{HS} = 1.5$$

respectively.

For the episodes of April 7, 1973, and August 2, 1973, already mentioned by Fronza et al. (1982) the results are summarized by the cost-effectiveness curves in Figures 1 and 2 (see also Bolzern and Fronza, 1981). The effectiveness index ϵ is defined as follows. Let $x_M^c(k)$ and $x_M^{nc}(k)$ denote the maximum ground-level concentration in the kth hour of the episode, with and without emission control, respectively. Next, consider the time averages

$$x_M^c = \sum_k x_M^c(k)/D; \qquad x_M^{nc} = \sum_k x_M^{nc}(k)/D$$

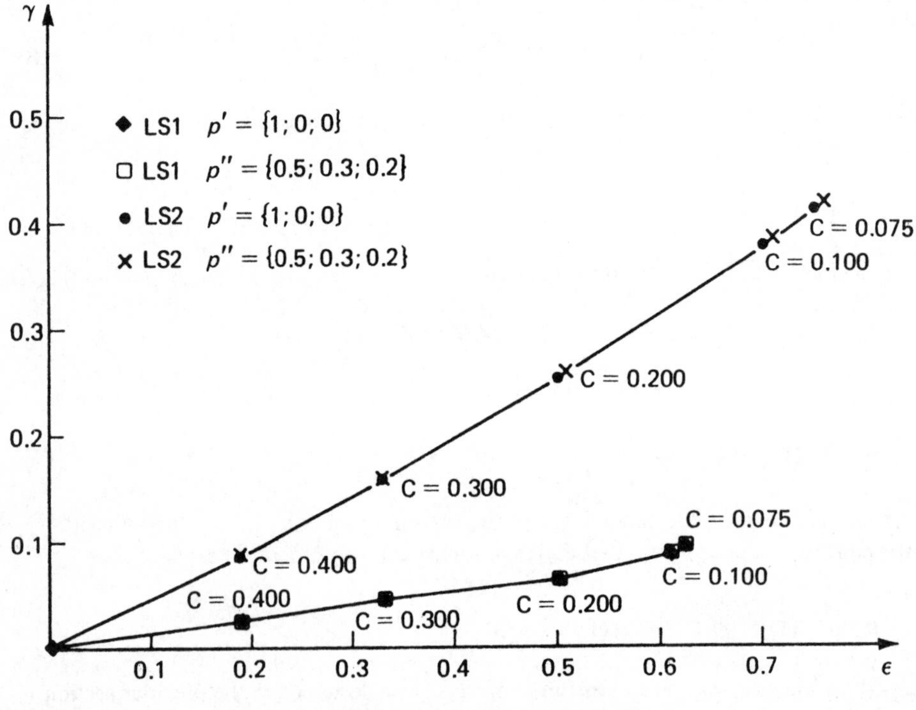

FIGURE 1 Cost-effectiveness curves for the episode of April 7, 1973.

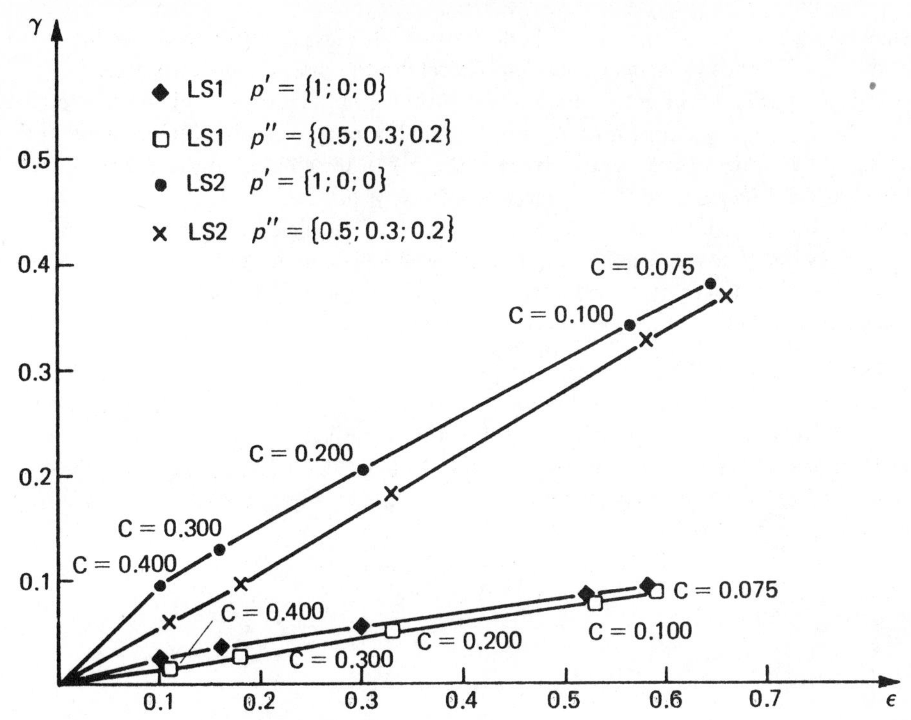

FIGURE 2 Cost-effectiveness curves for the episode of August 2, 1973.

where the sums are extended over all episode hours and D is the episode duration (in hours). The effectiveness index is then defined as

$$\epsilon = [x_{\mathrm{M}}^{\mathrm{nc}} - x_{\mathrm{M}}^{\mathrm{c}}]/x_{\mathrm{M}}^{\mathrm{nc}}$$

The cost index γ is the overall percent extra cost due to the control action, i.e.

$$\gamma = \left\{\sum_k [C^{\mathrm{c}}(k) - C^{\mathrm{nc}}(k)]\right\} \Big/ \sum_k C^{\mathrm{nc}}(k)$$

If eqns. (7) and (8) are taken into account, γ can simply be written as

$$\gamma = \left[\sum_k \gamma(k) Q(k)\right] \Big/ \sum_k Q(k)$$

where $\gamma(k)$ is given by eqn. (9).

From direction inspection of Figures 1 and 2 we can draw the following conclusions.

(a) In Figure 1 there is no difference between the results of control action under the two different triplets of weights p' and p''. Since $p' = \{1;0;0\}$ [control based only on 2-h forecast; see eqn. (2)], there is no (or, at least, seldom) significant improvement on considering also the 3-h and 4-h predictions to establish the control action. In contrast, these predictions yield some cost improvement in the case of Figure 2.

(b) For each c the effectiveness of the control action is mostly the same irrespective of which LS fuel is used although control by LS2 is more expensive. The only difference is at the upper bound of effectiveness (the end of the abscissa of the curves in Figures 1 and 2). Specifically, the less clean and less expensive fuel LS1 is generally sufficient for obtaining effectiveness values up to about 60%, but an effectiveness up to about 70% can only be reached by using the costly fuel LS2. Hence the use of a fuel like LS2, which is more than three times cleaner than LS1 [see eqn. (10)], allows one to obtain a maximum effectiveness which is only a few percent higher.

To complete the picture, some episode time patterns (for various c values, p weights and types of fuel) are shown in Figures 3 and 4. The corresponding time patterns of the control action $\alpha(k)$ are shown in Figures 5 and 6. Note particularly that the action is found to have an acceptably regular time profile; i.e., it does not consist of a series of undesirable control "shocks".

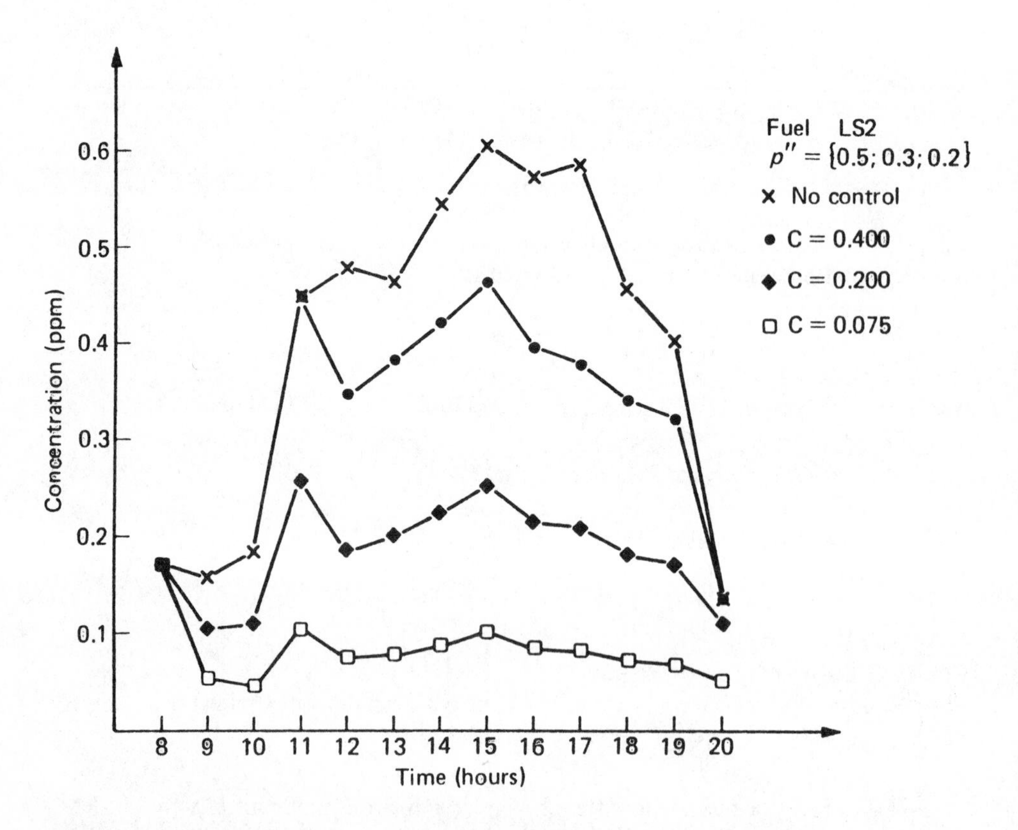

FIGURE 3 Concentration vs. time patterns for various control policies for station 9 and the episode of April 7, 1973.

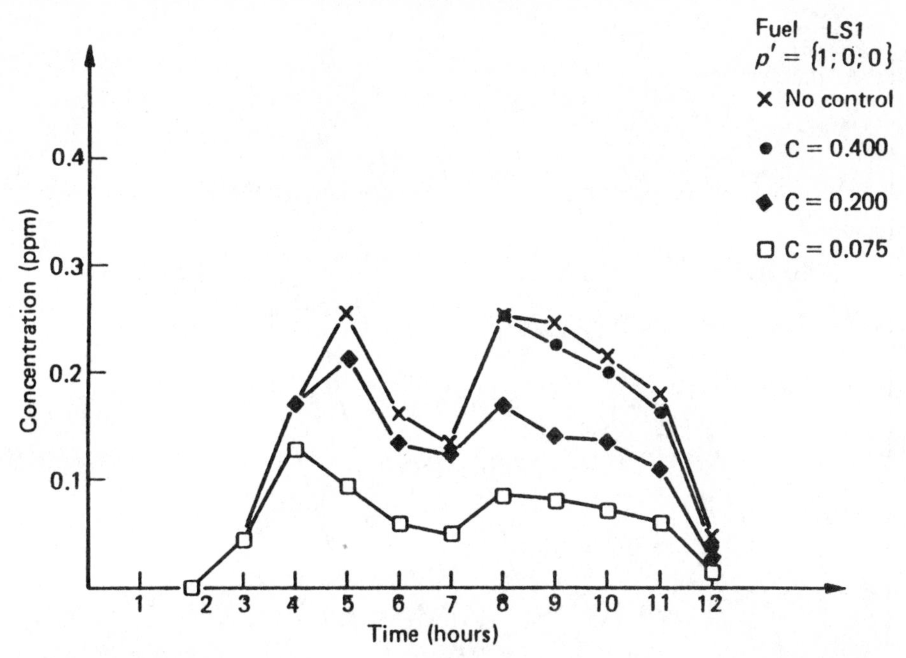

FIGURE 4 Concentration vs. time patterns for various control policies for station 30 and the episode of August 2, 1973.

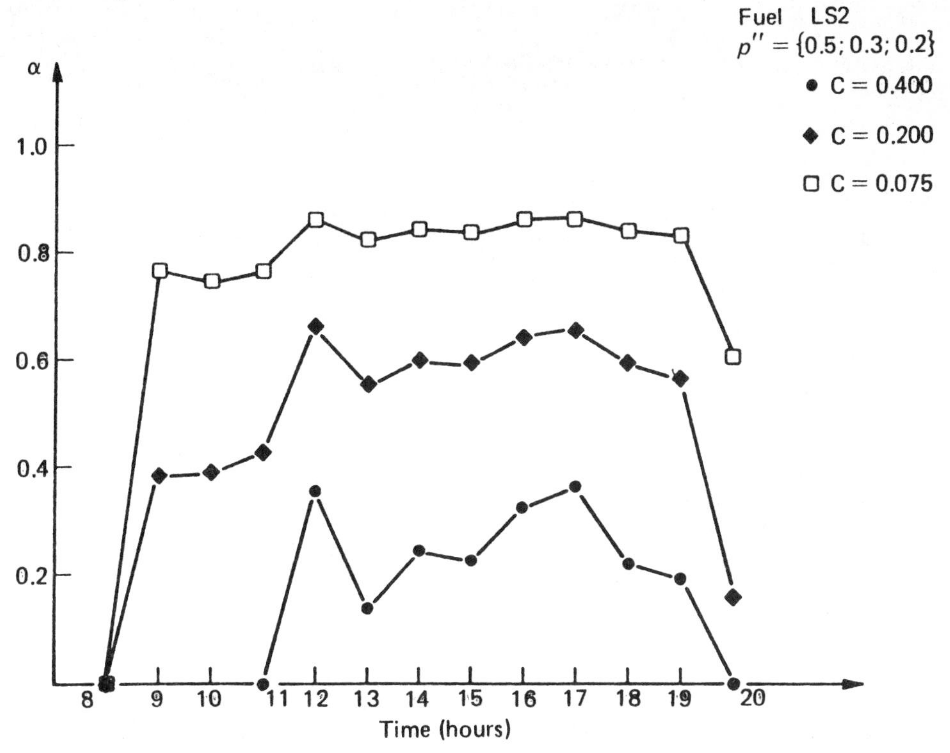

FIGURE 5 α vs. time patterns for the episode of April 7, 1973.

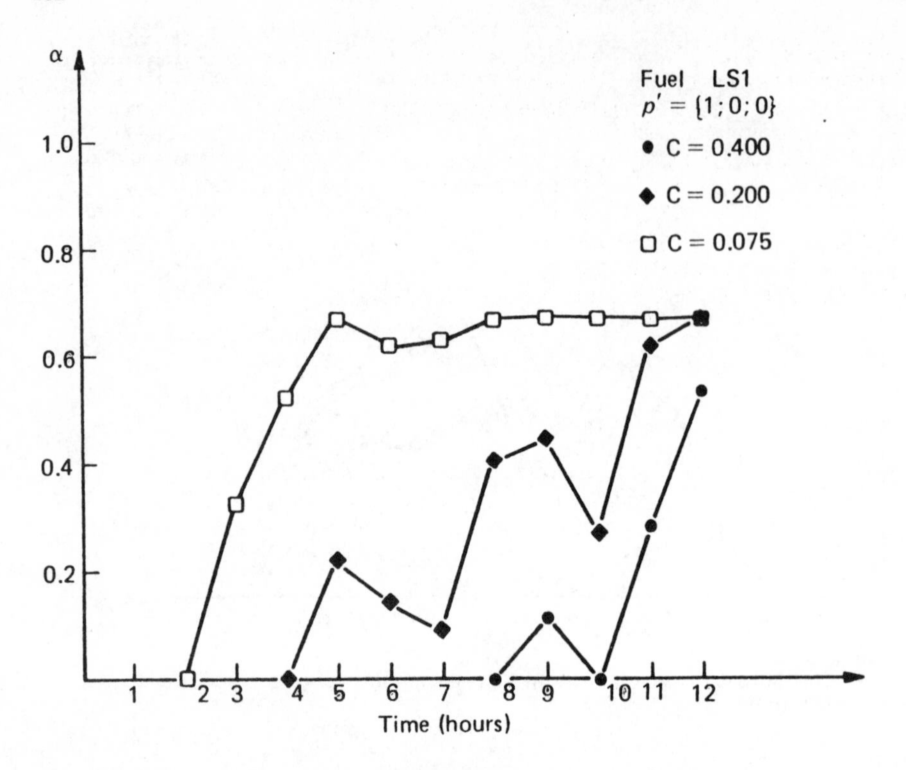

FIGURE 6 α vs. time patterns for the episode of August 2, 1973.

REFERENCES

Atkinson, S.E. and Lewis, D.H. (1974). A cost-effectiveness analysis of alternative air quality control strategies. J. Environ. Eng. Manag., 4:237–250.

Bankoff, S.G. and Hanzevack, E.L. (1975). The adaptive filtering transport model for prediction and control of pollutant concentration in an urban airshed. Atmos. Environ., 9:793–808.

Barbieri, D., Baroni, M., and Finzi, G. (1979). Multivariate real-time predictors of sulfur dioxide concentrations in Milan city. In Proc. NATO–CCMS Int. Tech. Meet. on Air Pollution Modeling, Rome, Italy.

Bolzern, P. and Fronza, G. (1981). Cost-effectiveness analysis of real-time control of SO_2 emission from a power plant. J. Environ. Manag. (in press).

Bonivento, C. and Tonielli, A. (1982). Short-term forecasting of local winds by black-box models. In G. Fronza and P. Melli (Editors), Mathematical Models for Planning and Controlling Air Quality. Pergamon Press, Oxford. (This volume, p. 139.)

Emanuel, W.R., Murphy, B.D., and Huff, D.D. (1978). Optimal siting of energy facilities for minimum air pollutant exposure on a regional scale. J. Environ. Manag., 7:147–155.

Finzi, G., Fronza, G., and Rinaldi, S. (1978). Stochastic modeling and forecast of the dosage area product. Atmos. Environ., 12:831–838.

Finzi, G., Zannetti, P., Fronza, G., and Rinaldi, S. (1979) Real-time prediction of SO_2 concentration in the Venetian Lagoon area. Atmos. Environ., 13:1249–1255.

Fronza, G., Spirito, A., and Tonielli, A. (1982). Kalman prediction of sulfur dioxide episodes. In G. Fronza and P. Melli (Editors), Mathematical Models for Planning and Controlling Air Quality. Pergamon Press, Oxford. (This volume, p. 161.)

Guldmann, J. and Shefer, D. (1976). Stack height as a means for air quality control: a mathematical programming approach. J. Environ. Manag., 4:241–249.

Leavitt, J.M., Carpenter, S.B., Blackwell, J.P., and Montgomery, T.L. (1971). Meteorological program for limiting power plant stack emissions. J. Air Pollut. Control Assoc., 21:400–405.

Runca, E., Melli, P., and Spirito, A. (1982). A *K* model for simulating the dispersion of sulfur dioxide in an airshed. In G. Fronza and P. Melli (Editors), Mathematical Models for Planning and Controlling Air Quality. Pergamon Press, Oxford. (This volume, p. 147.)

Seinfeld, J.H. and Kyan, C.P. (1971). Determination of optimal air pollution control strategies. Socio-Economic Planning Sci., 5:173–190.

Shepard, D.S. (1970). A load shifting model for air pollution control in the electric power industry. J. Air Pollut. Control Assoc., 20:756–761.

ARMAX STOCHASTIC MODELS OF AIR POLLUTION: THREE CASE STUDIES

P. Bolzern, G. Finzi, G. Fronza, and A. Spirito
Centro Teoria dei Sistemi, Milan (Italy)

1 INTRODUCTION

The absence of physical inputs (meteorological and emission variables) usually prevents black-box stochastic models of the AutoRegressive Integrated Moving Average (ARIMA) type [see for instance Box and Jenkins (1970) and Soeda and Sawaragi (1982)] from supplying an accurate description of pollutant dispersion in episode situations. In particular, the real-time ground-level concentration predictors derived from ARIMA models typically exhibit a delay effect when forecasting rapid concentration increases and subsequent decreases.

A better real-time episode predictor can be derived from the so-called ARIMAX (ARIMA with eXogenous inputs) stochastic mathematical representations [see for instance Box and Jenkins (1970)]. In such models pollutant concentrations at a certain instant are expressed as linear combinations of previous concentrations plus a linear combination of present and previous physical inputs plus noise terms.

In the present paper we summarize three case studies of the application of ARIMAX forecasting: a single-stack case (Section 2), an industrial-area case (Section 3) and an urban-area case (Section 4). The time scales and the variables taken as pollution representatives differ from case study to case study (in particular, the unit time interval ranges from 1 h to 1 day), as do the ARIMAX models themselves. In fact they are all Auto-Regressive Moving Average (with eXogenous inputs) (ARMAX) models since no time differentiation (Box and Jenkins, 1970) has proved to be useful.

The general conclusion is that in two cases forecast reliabilities ranging from 60 to 90% are reached while the result in the remaining case is not so satisfactory. Moreover, in all cases the forecast performance represents a significant improvement over trivial methods of prediction (such as setting future concentration equal to present concentration or other similar procedures).

2 THE POWER-PLANT CASE

In this section we summarize the work by Bacci et al. (1981) which aimed at an accurate forecast of summer SO_2 fumigation phenomena around the power plant at

Ostiglia in Northern Italy. Earlier ARMAX modeling (on a daily time scale) for the same case can be found in Finzi et al. (1978a, b).

The area is shown in Figure 1 together with the polluting source PP and the monitoring network (most of the sensors are located in the sector Ω of the prevailing pollutant fallout).

The variable taken as pollution representative is the 2-h Dosage Area Product (DAP) (Duckworth and Kupchanko, 1967) over sector Ω, i.e. the integral of the 2-h dosage over sector Ω.

The dynamics of this variable during the daylight period of each summer day were described by the following ARMAX model:

$$\mathrm{DAP}(k+1) = \alpha\{s(k+1), d(k+1)\}\,\mathrm{DAP}(k) + \beta\{s(k+1), d(k+1)\}\,p(k+1) + w(k) \tag{1}$$

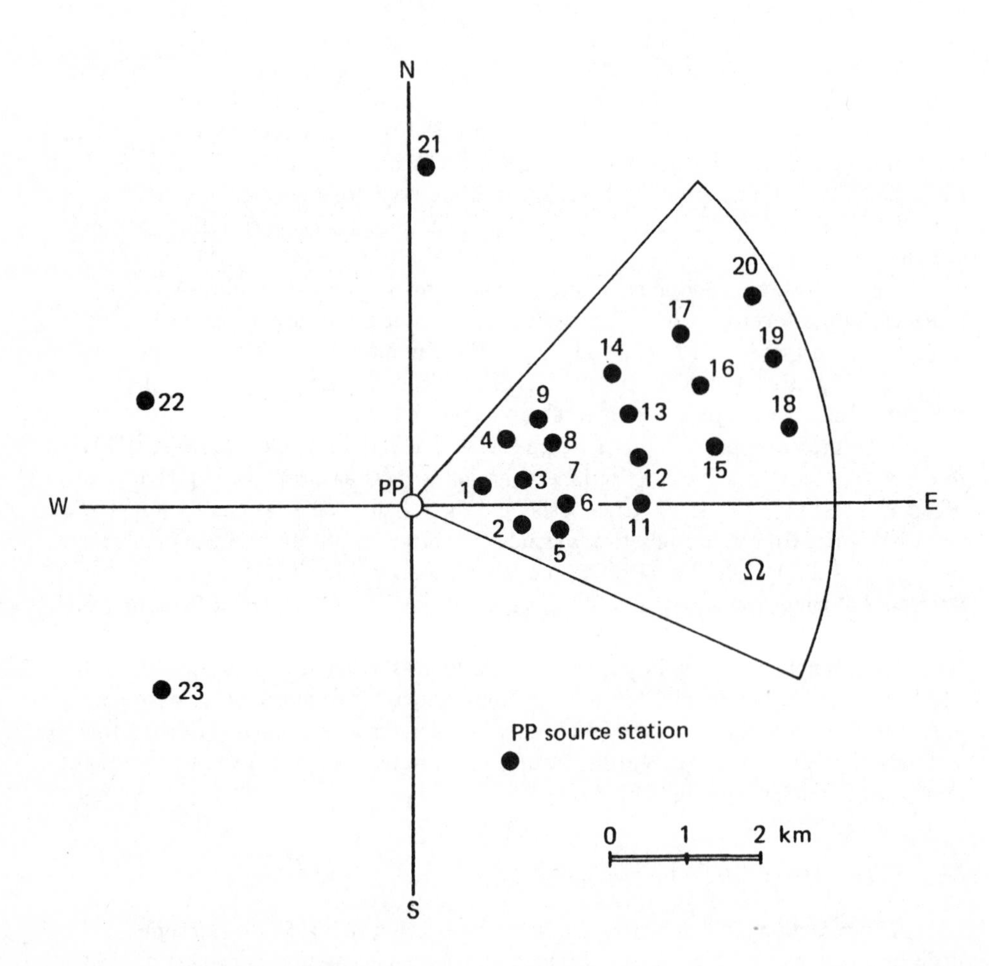

FIGURE 1 The power-plant case.

where $\Delta t = 2$ h. $p(k)$ is the average power generated during the kth time interval, i.e. during $\{(k-1)\Delta t, k\Delta t\}$ (this is an indirect measure of the overall emission during this interval). $s(k)$ is the properly defined atmospheric stability class which depends on the total radiation since sunrise, the average wind speed in the last 6 h, and the strength of the inversion at the end of the previous night [as measured by Pasquill's category (Pasquill, 1971)]. $d(k)$ is the wind direction class [$d(k) = 0$ if the wind does not blow toward Ω; $d(k) = 1$ otherwise]. α and β are model parameters depending on the already-mentioned meteorological variables. $\{w(k)\}_k$ is a stochastic process. Since $\{w(k)\}_k$ was found to be zero-mean white noise, the real-time predictor derived from model (1) is

$$\hat{\text{DAP}}(k+1|k) = \alpha\{\hat{s}(k+1|k), \hat{d}(k+1|k)\}\text{DAP}(k)$$

$$+\beta\{\hat{s}(k+1|k), \hat{d}(k+1|k)\}p(k+1) \tag{2a}$$

$$\hat{\text{DAP}}(k+2|k) = \alpha\{\hat{s}(k+2|k), \hat{d}(k+2|k)\}\hat{\text{DAP}}(k+1|k)$$

$$+\beta\{\hat{s}(k+2|k), \hat{d}(k+2|k)\}p(k+2) \tag{2b}$$

where $\hat{\text{DAP}}(k+r|k)$ is the forecast of DAP$(k+r)$ made at time $k\Delta t$; $\hat{s}(k+r|k)$ and $\hat{d}(k+r|k)$ are, respectively, the $(k+r)$th stability class and wind-direction forecasts made at $k\Delta t$; and $r = 1,2$.

The meteorological forecasts $\hat{s}(k+2|k)$ and $\hat{d}(k+2|k)$ were obtained by separate predictions. To be precise, $\hat{s}(k+r|k)$ was determined by assuming a persistent wind speed and by applying a radiation predictor derived from experimental evidence, while $\hat{d}(k+r|k)$ was obtained by probabilistic considerations involving the current measure $d(k)$ and the hour of the day corresponding to $\{(k-1)\Delta t, k\Delta t\}$.

The forecast performance of the DAP predictor (2) is shown in the first two columns of Table 1 where θ is the correlation between the predicted and the measured DAP [actually the DAP evaluated by weighting the point-concentration measurements using the polygons method (see, for instance, Gray, 1974)], and θ_F is the correlation between the predicted and the measured DAP in fumigation situations.

Table 1 also shows how much is lost in DAP forecasting because of inaccurate meteorological prediction. This is illustrated by the last two columns of the table, which correspond to running predictor (2) in correspondence with $\hat{s}(k+r|k) = s(k+r)$, $\hat{d}(k+r|k) = d(k+r)$, i.e. under conditions of perfect forecasting of future meteorology. The loss is not negligible but it is not as marked as one might have feared in view of the gross nature of the meteorological predictors.

TABLE 1 Performance of the 2-h and 4-h DAP predictors in the power-plant case.

	Predictor			
Forecast quality	2 h ahead (forecast meteorology)	4 h ahead (forecast meteorology)	2 h ahead (true meteorology)	4 h ahead (true meteorology)
θ	0.87	0.76	0.91	0.84
θ_F	0.72	0.60	0.80	0.68

3 THE INDUSTRIAL-AREA CASE

This section is devoted to a description of the ARMAX modeling developed by Finzi et al. (1979) for the Venetian lagoon. The area is shown in Figure 2. It was divided into three subregions, roughly corresponding to the urban centers of Venice, Mestre, and Marghera–Porto Marghera. The industrial sources are located in Porto Marghera (see also Runca et al., 1982).

The following ARMAX model of the hourly summer SO_2 DAP was used:

$$\ln\{\mathrm{DAP}_j(k+1)\} - \mu_j = \phi_j[\ln\{\mathrm{DAP}_j(k)\} - \mu_j] + \psi_{1j}[\ln\{v_j(k+1)\} - \mu_{vj}] + \psi_{2j}[\ln\{d_j(k+1)\} - \mu_{dj}] + \epsilon(k) \qquad j = 1,2,3 \qquad (3)$$

where $\mathrm{DAP}_j(k)$ is the kth hour DAP in subregion j, μ_j is the mean of $[\ln\{\mathrm{DAP}_j(k)\}]_k$, $v_j(k)$ is the average wind speed blowing from the sources towards subregion j in the H_j hours before the kth hour, μ_{vj} is the mean of $[\ln\{v_j(k)\}]_k$, $d_j(k)$ is the percentage of time (in the last H_j hours) during which the wind blew from the sources towards subregion j, μ_{dj} is the mean of $[\ln\{d_j(k)\}]_k$, ϕ_j, ψ_{1j}, ψ_{2j} are model parameters, and $\{\epsilon(k)\}_k$ is a stochastic process.

By model calibration the most suitable H values were found to be $H_1 = 2$ (for Venice), $H_2 = 1$ (for Mestre), and $H_3 = 1$ (for Marghera–Porto Marghera). Moreover $[\epsilon(k)]_k$ was found to be zero-mean white noise.

The 1-h and 4-h forecast performances of the predictor derived from model (3) are shown, respectively, in Tables 2 and 3 (first two columns) (only the forecast quality for the DAP in Marghera, the most-polluted area, and for Venice, the least-polluted subregion, are supplied for brevity). Here S is the standard deviation of the forecast error, S^e is the standard deviation of the forecast error in "episode" situations, the upper bound is the performance of the DAP predictor for a perfect meteorological forecast, and the lower bound is the performance of the DAP predictor for a persistent meteorological forecast.

A comparison with the quality of the trivial persistence DAP predictor can be made by considering the third columns in Tables 2 and 3. Clearly the ARMAX 1-h prediction is

TABLE 2 Performance of 1-h DAP predictors for the Marghera and Venice subareas.

	Predictor			
	ARMAX			
Forecast quality	Upper bound	Lower bound	Persistence	Cyclostationary ARMA
Marghera				
θ	0.74	0.74	0.73	0.77
S(ppb × h)	0.20	0.21	0.22	0.20
S^e(ppb × h)	0.43	0.43	0.44	0.42
Venice				
θ	0.81	0.79	0.78	0.83
S(ppb × h)	6	6	7	5
S^e(ppb × h)	0.18	0.19	0.20	0.17

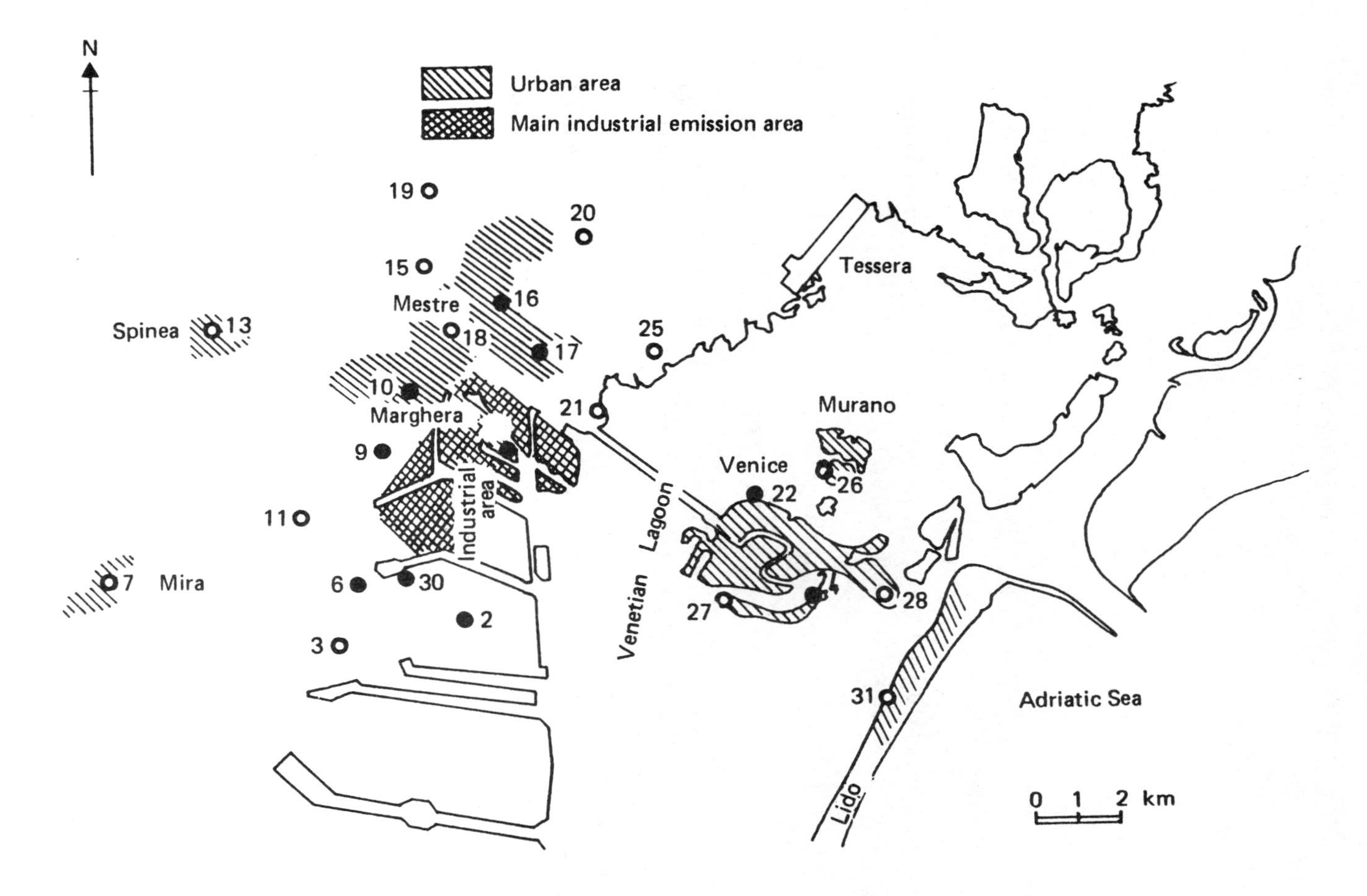

FIGURE 2 The Venetian case.

TABLE 3 Performance of 4-h DAP predictors for the Marghera and Venice subareas.

	Predictor			
	ARMAX			
Forecast quality	Upper bound	Lower bound	Persistence	Cyclostationary ARMA
Marghera				
θ	0.45	0.35	0.29	0.50
S(ppb × h)	0.27	0.28	0.36	0.25
S^e(ppb × h)	0.58	0.63	0.68	0.53
Venice				
θ	0.58	0.45	0.33	0.60
S(ppb × h)	8	9	12	7
S^e(ppb × h)	24	27	33	23

substantially equivalent to the trivial prediction while the 4-h ARMAX prediction is much better, although still not very satisfactory.

A better performance can be obtained (see the fourth columns of Tables 2 and 3) by using the predictor derived from the following cyclostationary (or periodic) ARMA (ARMAX without any exogenous input) model:

$$\ln\{\mathrm{DAP}_j(24q + t + 1)\} - \mu_{(t+1)j} = \phi_{tj}[\ln\{\mathrm{DAP}_j(24q + t)\} - \mu_{tj}] + \epsilon_j(24q + t) \quad (4)$$

where $t = 1,2,\ldots,24$ is the hour-of-day index, $q = 0,1,\ldots$ is the day index, $\mathrm{DAP}_j(24q + t)$ is the DAP in the jth subregion in the tth hour of the qth day, μ_{tj} is the mean of $[\ln\{\mathrm{DAP}_j(24q + t)\}]_q$, ϕ_{tj} is the model parameter, and $\{\epsilon(24q + t)\}_{q,t}$ is a stochastic process.

The improvement in the performance of the predictor derived from model (4) compared to the ARMAX predictor derived from model (3) can be explained only by the existence of daily cycles of emission which are indirectly accounted for by parameter periodicity in model (4) but which are not taken into account by model (3).

However, in general the performance cannot yet be regarded as very satisfactory, so complex predictors of a different nature have been considered (Fronza et al., 1982).

4 THE URBAN CASE

In this section we summarize the recent work of Finzi et al. (1980) on the Milan metropolitan area which extends their earlier work (Finzi et al., 1977).

The region under consideration is shown in Figure 3, together with its subdivision into three sectors, R_1, R_2, and R_3, and the network of ten SO_2-monitoring stations. Winter daily SO_2 pollution, which is almost entirely due to residential heating, was analyzed. In particular, the vector of the three daily SO_2 DAPs was considered and its dynamics were described by means of the two following multivariate ARMAX models.

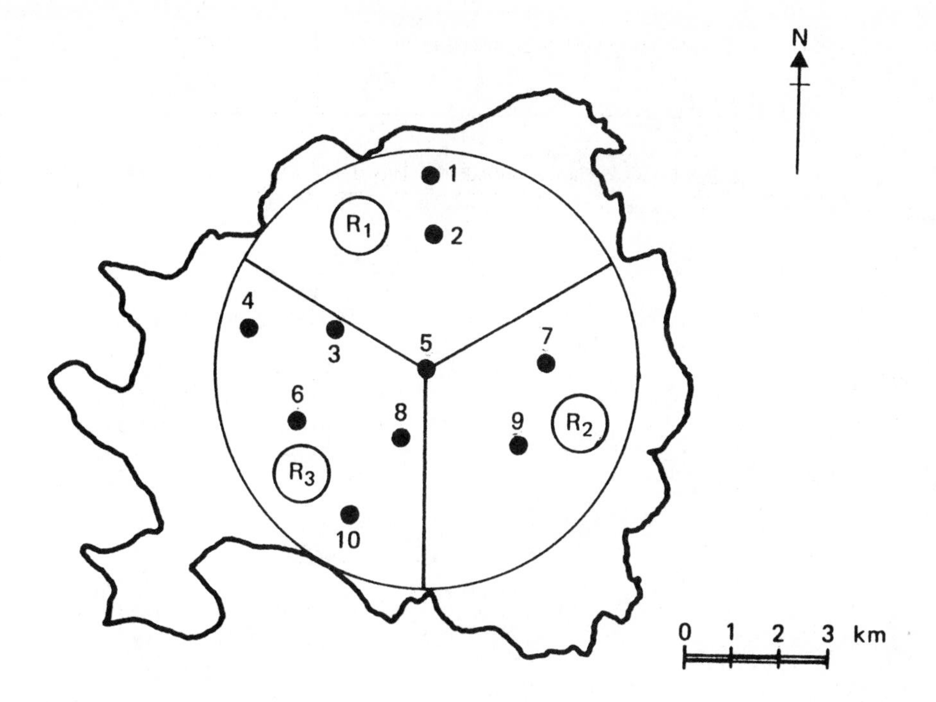

FIGURE 3 The Milan case.

4.1 Single-Input Model

This model has a temperature input introduced as an emission-representative variable (see also Bolzern et al., 1981):

$$\mathrm{DAP}(k+1) = \phi \mathrm{DAP}(k) + f\{T(k)\} + \eta(k) \tag{5}$$

where DAP(k) is the DAP vector on the kth day, $T(k)$ is the average temperature on the kth day, $f\{T(k)\} = |f_j\{T(k)\}|$ where $f_j\{T(k)\} = a_j/b_j + T(k)$ for $j = 1,2,3$, $\{\eta(k)\}_k$ is a three-variate stochastic process, and ϕ, a_j, b_j are model parameters. The performance of the 1-day predictor derived from model (5) is shown in Table 4 where S_j is the mean DAP and the superscript corresponds to episode situations. The performance should be compared with that achieved using the trivial persistence predictor (Table 4, last column).

4.2 Two-Input Model

This model has a further input, namely the wind speed:

$$\mathrm{DAP}(k+1) = \phi \mathrm{DAP}(k) + f\{T(k)\} + \psi v(k+1) + \eta(k) \tag{6}$$

TABLE 4 Performance of 1-day-ahead DAP predictors for subregions R_1, R_2, and R_3.

	Predictor			
		ARMAX (inputs $T(k)$, $v(k+1)$)		
Forecast quality	ARMAX (input $T(k)$)	Upper bound	Lower bound	Persistence
Subregion R_1				
θ	0.76	0.84	0.77	0.68
θ^e	0.61	0.69	0.62	0.53
S/μ	0.36	0.32	0.36	0.44
S^e/μ	0.34	0.28	0.33	0.36
Subregion R_2				
θ	0.79	0.84	0.80	0.73
θ^e	0.66	0.72	0.67	0.65
S/μ	0.35	0.31	0.34	0.40
S^e/μ	0.31	0.27	0.29	0.31
Subregion R_3				
θ	0.75	0.82	0.77	0.69
θ^e	0.42	0.50	0.44	0.33
S/μ	0.36	0.31	0.35	0.41
S^e/μ	0.32	0.28	0.31	0.37

where $v(k+1)$ is the average wind speed on the $(k+1)$th day and ψ is the vector of the parameters. The upper bound (for a perfect wind-speed forecast) and the lower bound (for a persistent wind-speed forecast) of the forecast performance are also shown in Table 4.

It should be particularly noted that, unlike the Venetian case, both ARMAX predictors not only supply a marked improvement with respect to the trivial DAP forecast but also exhibit a satisfactory forecast quality.

REFERENCES

Bacci, P., Bolzern, P., and Fronza, G. (1981). A stochastic predictor of air pollution based on short-term meteorological forecasts. J. Appl. Meteorol., 20: 121–129.

Bolzern, P., Fronza, G., Runca, E., and Überhuber, C. (1981). Statistical analysis of winter sulphur dioxide concentration data in Vienna. Atmos. Environ., (in press).

Box, G.E.P. and Jenkins, G.M. (1970). Time Series Analysis, Forecasting and Control. Holden–Day, San Francisco, California.

Duckworth, S. and Kupchanko, E. (1967). Air analysis, the standard Dosage Area Product. J. Air Pollut. Control Assoc., 17:379–387.

Finzi, G., Fronza, G., and Spirito, A. (1977). Univariate stochastic models and real-time predictors of daily SO_2 pollution in Milan. In Proc. NATO–CCMS Int. Tech. Meet., Air Pollution Modeling, 8th, Louvain-la-Neuve, Belgium.

Finzi, G., Fronza, G., and Rinaldi, S. (1978a). Stochastic modeling and forecast of the dosage area product. Atmos. Environ., 12:831–838.

Finzi, G., Fronza, G., Rinaldi, S., and Spirito, A. (1978b). Prediction and real-time control of SO_2 pollution from a power plant. In Proc. APCA Annu. Meet., Houston, Texas.

Finzi, G., Zannetti, P., Fronza, G., and Rinaldi, S. (1979). Real-time prediction of SO_2 concentration in the Venetian Lagoon area. Atmos. Environ., 13:1249–1255.

Finzi, G., Fronza, G., and Spirito, A. (1980). Multivariate stochastic models of sulphur dioxide pollution in an urban area. J. Air Pollut. Control Assoc., 30: 1212–1215.

Gray, D.M. (Editor) (1974). Handbook on the Principles of Hydrology. Water Information Center, Inc.

Fronza, G., Spirito, A., and Tonielli, A. (1982). Kalman prediction of sulfur dioxide episodes. In G. Fronza and P. Melli (Editors), Mathematical Models for Planning and Controlling Air Quality. Pergamon Press, Oxford. (This volume, p. 161.)

Pasquill, F. (1971). Atmospheric Diffusion. Van Nostrand, New York.

Runca, E., Melli, P., and Spirito, A. (1982). A K model for simulating the dispersion of sulfur dioxide in an airshed. In G. Fronza and P. Melli (Editors), Mathematical Models for Planning and Controlling Air Quality. Pergamon Press, Oxford. (This volume, p. 147.)

Soeda, T. and Sawaragi, Y. (1982). ARIMA and GMDH forecasts of air quality. In G. Fronza and P. Melli (Editors), Mathematical Models for Planning and Controlling Air Quality. Pergamon Press, Oxford. (This volume, p. 195.)

ARIMA AND GMDH FORECASTS OF AIR QUALITY

T. Soeda
Tokushima University, Tokushima (Japan)

Y. Sawaragi
Kyoto University, Kyoto (Japan)

1 INTRODUCTION

The problem of forecasting and controlling air-pollutant concentrations is generally considered to be important because of the effects of pollution on human health. In order to obtain a reliable forecast (and subsequently to take control action when the forecast pollution exceeds a certain level) models describing the dynamics of pollutant concentration are required.

These mathematical representations are generally classified as physical and non-physical. The former models consist of diffusion equations and describe both the temporal and the spatial distribution of the pollutant (Desalu et al., 1974; Kondo, 1975), while the latter models are usually referred to as "time-series models" (Box and Jenkins, 1976; Soeda and Ishihara, 1974; Akizuki and Shirai, 1975; McCollister and Wilson, 1975; Ishihara and Soeda, 1976; Sawaragi et al., 1976).

The purpose of this paper is to compare the performances of various non-physical models in forecasting pollution levels and to illustrate new forecasting techniques. First, the accuracy of forecasting of pollution levels by four time-series models is evaluated (for three kinds of performance index). Secondly, the multiple linear-regression model is revised by taking into consideration wind and other meteorological variables, and the forecasting improvement depending on these factors is evaluated. Thirdly, the accuracy of time-series models is discussed for the required data length, and the confidence intervals of forecasts at a fixed time point are evaluated. Furthermore, the Kalman filtering technique (Meditch, 1969) which considers the effects of input and measurement noises is applied to the forecasting problem. Fourthly, forecasting based on the Group Method of Data Handling (GMDH) is proposed and the accuracy using this technique is compared with the results obtained using time-series models. In this case data-processing and modeling techniques using an Adaptive Digital Filter (ADF) derived from the output-error structure are compared with the GMDH method. The approaches mentioned here are discussed using numerical examples based on data for Japan and Italy.

2 COMPARISON OF THE ACCURACY OF FORECASTING OF POLLUTION LEVELS BY TIME-SERIES MODELS

The forecasting accuracy of four non-physical models is evaluated by the following three performance indices:

$$J_M = \sum_{i=1}^{N} \sum_{k=1}^{24} [x_i(k) - \hat{x}_i\{k/(k-m)\}]^2 / 24$$

$$J_N = \sum_{i=1}^{N} \left(\sum_{k=1}^{24} [x_i(k) - \hat{x}_i\{k/(k-m)\}]^2 / \sum_{k=1}^{24} x_i^2(k) \right)$$

$$J_I = \sum_{i=1}^{N} \left(\sum_{k=1}^{24} [x_i(k) - \hat{x}_i\{k/(k-m)\}]^2 / \sum_{k=1}^{24} [x_i^2(k) + \hat{x}_i^2\{k/(k-m)\}] \right)$$

where $x_i(k)$ denotes the pollution concentration in the kth hour of the ith day, $\hat{x}_i\{k/(k-m)\}$ denotes the forecast of $x_i(k)$ made at the time $k-m$, and N is the number of days taken into account. The first index is simply the sum of the squares of the forecasting errors. The second and third indices represent the forecasting errors, normalized according to pollution levels and according to pollution levels plus the forecasts, respectively. The third index was suggested by A.G. Ivakhnenko (at the 4th IFAC Symposium on Identification and Parameter Estimation, Tbilisi, USSR, 1976).

2.1 AutoRegressive (AR) Model

In the fitting of a nonstationary time series, obtained through a sequence of hourly-measured pollution levels, by means of an AR model, first the nonstationarity or trend of the data must be removed. The resulting stationary sequence is then fitted to an AR model as follows.

(i) In order to obtain the stationary time series from the observed data sequence $\{x(k)\}$ we subtract the value of the moving average at each time k from the original data sequence. We let the new time series be $\{z(k)\} = \{x(k) - \bar{x}(k)\}$.

(ii) When the new time series $\{z(k)\}$ is not regarded as stationary we try to take the difference of the successive values.

(iii) For the resulting stationary data the order of the AR model is determined by the FPE method (Akaike, 1969, 1970).

2.2 Multiple Linear-Regression Model

Assuming that the measured data on pollution levels have a 24-h daily cycle, we consider the following multiple linear-regression model:

$$\boldsymbol{x}(k) = A(k, k-1)\boldsymbol{x}(k-1) + \boldsymbol{b}(k-1)$$

where $\boldsymbol{x}(k)$ denotes an $n \times 1$ vector representing pollution levels and meteorological factors (e.g. wind, temperature, humidity, and cloud) which have strong effects on pollution levels. These factors are chosen by computing their correlation with the pollution levels. The time-varying matrix $A(k,k-1)$ and deterministic input $\boldsymbol{b}(k-1)$ are determined by processing the measured data in the sense of the ensemble average:

$$A(k,k-1) = E\{\tilde{\boldsymbol{x}}(k)\tilde{\boldsymbol{x}}^{\mathrm{T}}(k-1)\}E\{\tilde{\boldsymbol{x}}(k-1)\tilde{\boldsymbol{x}}^{\mathrm{T}}(k-1)\}^{-1}$$

$$\boldsymbol{b}(k-1) = E\{\boldsymbol{x}(k)\} - A(k,k-1)E\{\boldsymbol{x}(k-1)\}$$

where E denotes the ensemble average with respect to the measured data at each sampling time, T is the vector-transformation superscript, and $\tilde{\boldsymbol{x}}(k)$ denotes $\boldsymbol{x}(k) - E\{\boldsymbol{x}(k)\}$. The forecast $\hat{\boldsymbol{x}}\{k/(k-m)\}$ is given by

$$\hat{\boldsymbol{x}}\{k/(k-m)\} = A(k,k-m)\boldsymbol{x}(k-m) + \sum_{j=k-m+1}^{k} A(k,j)\boldsymbol{b}(j-1)$$

where

$$A(k,k) = I$$

$$A(k,j) = A(k,k-1)A(k-1,k-2)\cdots A(j+1,j) \qquad (\text{for } j<k)$$

2.3 Box–Jenkins and Persistence Models

The Box–Jenkins model is called the AutoRegressive Integrated Moving Average (ARIMA) model (Box and Jenkins, 1976). If the AR operator is of order p, then the dth difference is taken and the moving average (MA) operator is of order q; the model is labeled ARIMA(p,d,q).

The method based on the principle of persistence (McCollister and Wilson, 1975) is often used for short-range weather forecasting. The principle consists in assuming that the pollution levels in the next few hours will be the same as the present levels.

2.4 Case Study

Here we compare the accuracy of pollution-level forecasting by the four time-series models presented above. The comparison is carried out by evaluating the three performance indices defined earlier.

The recorded data for the case study consist of hourly measurements in Tokyo from June to December, 1971, and in Tokushima from May to July, 1975. The sensors in Tokyo and Tokushima are located in the center of each city. The pollution levels (pphm) of O_x, NO, NO_2, CO, and SO_2 were measured by counting 0.5 pphm. A wind speed below 0.1 m/s was considered as zero, and the wind direction was measured in units of 22.5° starting from the north. The weather was classified as fine, cloudy, or rainy.

Table 1 shows the forecasting accuracy for SO_2 in Tokushima obtained by using the four time-series models, as measured by the three performance indices. For the Box–Jenkins model it was assumed that $d = 1$ and $q = 0$; i.e. the time-series model preprocessed by the first difference of the measurement sequence was used. The multiple linear-regression model was built for the forecasting of five pollution data: concentrations of O_x, NO, NO_2, CO, and SO_2. Table 1 illustrates that the AR model is the best in 1-h forecasting while the multiple linear-regression model is the best in 2-h and 3-h forecasting. From the evaluation of J_N and J_I, the 1-h and 2-h forecasting accuracy of the AR model is found to be very close to that of the multiple linear-regression model because of the normalization of the forecasting errors at high pollution levels.

TABLE 1 Forecasting accuracy of time-series models.

Forecast	Performance index	AR model	Multiple linear regressive model	Box–Jenkins model (ARI)	Persistence model
1 h in advance	J_M	2,602	2,662	2,664	3,168
	J_N	1.05	1.07	1.06	1.17
	J_I	0.52	0.53	0.53	0.59
2 h in advance	J_M	5,785	5,650	7,244	8,711
	J_N	2.69	2.60	2.80	3.08
	J_I	1.25	1.24	1.38	1.54
3 h in advance	J_M	9,312	9,029	12,323	15,157
	J_N	4.27	4.01	4.50	5.21
	J_I	1.91	1.86	2.26	2.61

Table 2 shows the 1-h forecasting accuracy for SO_2 in July at Tokushima in four cases using linear-regression models. The wind velocity was chosen as a state variable because it is considered to have a significant correlation with the pollution levels. Other meteorological factors such as temperature, humidity, visibility, cloud, etc., are treated by weather classification. The following four cases were distinguished:

case 1, the weather and the wind are considered;
case 2, the weather is ignored and the wind is considered;
case 3, the weather is considered and the wind is ignored;
case 4, the weather and the wind are both ignored.

TABLE 2 Forecasting accuracy of linear-regression models in four cases.

Case	J_M	J_N	J_I
1	2221	0.891	0.450
2	2131	0.854	0.431
3	2013	0.811	0.401
4	2020	0.814	0.409

In cases 1 and 2 the wind direction is divided into easterly and southerly components. When the weather is taken into account, the SO_2 data are grouped into three sets and three models are built by processing the corresponding data. Case 4 gives a single linear-regression model. The numerical results in Table 2 show that the forecasting accuracy of the models that take the wind into account is not very high. However, the forecasting accuracy is slightly improved by the two models using weather classification, but this should not be taken as a general conclusion (because of the size of the present experiment). The accuracy and characteristics of the forecasts by the AR model and the regression model are compared in Table 3.

TABLE 3 Accuracy and characteristics of regression models.

Characteristic	ARMA model	Multiple linear regression model	ARI model	Persistence model
Method for removing the trend of the time series	Moving average	Regression analysis (ensemble average)	1st difference	Principle of persistence
Order of the model	Final prediction error	1st order	Final prediction error	
Parameters	Time invariant	Time variant	Time invariant	
Sum for the period June 1–15 (June 1–30)* of the mean square error (a)				
1 h in advance	1175 (1.06)	1104 (1.07)	1186 (1.06)	1593 (1.17)
2 h in advance	3511 (2.72)	2823 (2.60)	3610 (2.80)	4766 (3.08)
3 h in advance	6160 (4.55)	4511 (4.01)	6594 (4.50)	8600 (5.21)
Comparison of forecast with the daily maximum	Predicted value is higher than the daily maximum	Predicted value is lower than the daily maximum	Predicted value is higher than the daily maximum	

3 DETERMINATION OF THE DATA LENGTH FOR SETTING UP MODELS

In the previous section several time-series models were set up by processing measured pollution data, but no reference was made to the required data length. For this problem, the autocorrelation function and the FPE method are used here to check the acceptable data length, where a data length of 24 (h) $\times$ p(days) was considered for SO_2 data in Tokushima with p (10–60 by 10). Using the autocorrelation function we found the estimated data lengths to be similar to each other with $p \geqslant 30$. For the FPE method, however, the data length contains two parameters as model order and data length, and the stationarity and normality of data must be examined. It was found that the data are stationary by a run test and that their distribution can be regarded as normal by a χ^2 test

(Table 4). The FPE result with some fixed p (10–60 by 10) indicates similar profiles with $p \geqslant 30$. Another FPE result with fixed model orders (5 and 18) indicates that the required data length of 30 days is satisfactory by the principle of FPE minimum and that the model order 18 is an acceptable value.

TABLE 4 Normality and stationarity tests.

	Normality		Stationarity	
Time	Result	χ^2 [a]	Result	Run[b]
1:00	0	2.77	0	6
2:00	0	4.63	0	6
3:00	0	3.00	0	4
4:00	0	1.13	0	7
5:00	0	3.47	0	7
6:00	0	2.53	0	6
7:00	0	2.30	0	6
8:00	0	7.43	0	3
9:00	0	6.27	0	3
10:00	0	0.90	0	5
11:00	0	2.30	0	5
12:00	0	6.03	×	2
13:00	0	5.10	0	4
14:00	0	7.90	0	6
15:00	×	10.23	0	6
16:00	0	4.17	0	6
17:00	0	7.67	0	4
18:00	0	6.27	0	7
19:00	0	4.40	0	7
20:00	0	6.27	0	6
21:00	0	5.33	0	6
22:00	0	7.67	0	6
23:00	0	5.80	0	6
24:00	0	2.77	0	6

The data are for May and June, 1975, in Tokushima. The sample size at each time was 61.

[a] The χ^2 test of normality is $P(\chi^2 > \chi_{n;\alpha^2}) = \alpha$ where the significance level $\alpha = 0.05$ and the freedom $f = 4$ (the number of class = 7). In this class $\chi_{n;\alpha^2} = 9.49$. If $\chi^2 < 9.49$ the hypothesis is accepted (shown by symbol 0) and if $\chi^2 > 9.49$ the hypothesis is rejected (shown by symbol ×).

[b] The run test of stationarity is as follows. The 61 data at each time are divided into ten classes, and the standard deviation is calculated in each case. When the number of the run is between 2 and 9, the stationarity is accepted; when it is not, the stationarity is rejected.

4 CONFIDENCE INTERVAL OF THE FORECAST MEAN AT A FIXED TIME POINT

In this section the confidence intervals of the forecast means of pollution levels (at a fixed point) are estimated from past measured data; the following two situations are considered.

(i) All the pollution levels (O_x, NO, NO_2, SO_2, CO) are measured at a past time point $(k_f - m)$ where k_f denotes a fixed time point.
(ii) One pollution level (SO_2) is measured at five past measurement time points, $k_f - m_i$ $(i = 1,2,...,5)$ where $m_i > m_j$ for $i > j$.

Other combinations of the measured pollution levels and the measurement time points can be considered. The statistics for the pollution data can be derived by a method similar to that used in the second case. However, much more computation is needed for the experiment so the experiment was restricted to the two cases mentioned.

The confidence intervals were computed using Tokyo measurements. Table 5 shows the forecast means, the RMS values of some pollution levels, and the confidence intervals of the forecast means. Here the Student *t* distribution was applied and the level of significance was set at 0.1 (Bendat and Piersol, 1971). The use of the Student *t* distribution is justified by the fact that the pollution levels measured in Tokushima and Tokyo can be regarded as nearly normal. It can be seen from Table 5 how the forecasts changed according to the measurement time points and the factors measured.

TABLE 5 Forecast means, RMS values, and confidence intervals.

Factor to be predicted	Time	Measured factor	Measurement time	Predicted values		Confidence interval of predicted mean	Actual values	
				Mean	RMS		Mean	RMS
O_x	13	O_x	8,9,10,11,12	13.68	4.98	12.80–14.56	13.85	4.56
		NO_2		14.00	2.90	13.49–14.51		
		SO_2		13.55	4.28	13.44–13.66		
		O_x, NO, NO_2, CO, SO_2	8	13.54	0.64	13.43–13.65		
			10	13.83	0.96	13.66–14.00		
			12	14.03	2.44	13.60–14.46		
CO	13	O_x	8,9,10,11,12	2.69	3.23	2.12–3.26	2.81	1.65
		O_x, NO, NO_2, CO, SO_2	8	2.98	0.71	2.86–3.10		
SO_2	13	O_x, NO, NO_2, CO, SO_2	8	8.82	2.41	8.39–9.24	8.25	5.42
			10	9.10	3.24	8.52–9.68		
			12	8.53	4.40	7.75–9.31		

5 MODELING AND FORECASTING BY USING AN ADAPTIVE DIGITAL FILTER (ADF)

The schemes for modeling and forecasting for an environmental system by using an ADF are as follows.

In recent years, Widrow has proposed an ADF technique based on the Least Mean Square (LMS) algorithm (Widrow et al., 1975), and its extension has been reported by White (1975). The main advantage of this algorithm is its computational simplicity in the real-data processing.

5.1 Modeling and Forecasting

In this section an ADF identification technique utilizing the output-error structure is applied (Kikuchi et al., 1979). For time-series modeling the data are assumed to be derived by the random-shock sequence. The time-series model can therefore be identified using the idea that the random shock can be estimated as 1-h-ahead forecasting error.

We assume that the environmental system is represented as ARMA (p,q):

$$H(z) = (1 + \theta_1 z^{-1} + \theta_2 z^{-2} + \ldots + \theta_q z^{-q})/(1 - \phi_1 z^{-1} - \phi_2 z^{-2} - \ldots - \phi_p z^{-p})$$

where ϕ_k and θ_k are, respectively, AR and MA parameters and (p,q) are the orders of the AR and the MA parts, respectively. For the identification of ARMA(p,q), the ADF is built up as

$$G(z) = (b_1 + b_2 z^{-1} + \ldots + b_p z^{-p+1})/(1 - a_1 z^{-1} - a_2 z^{-2} - \ldots - a_q z^{-q})$$

The estimated transfer function $\hat{H}(z)$ of the environmental system $H(z)$ can then be obtained as

$$\hat{H}(z) = 1/\{1 - G(z)z^{-1}\}$$

Thus the AR and MA parameters can be estimated as follows:

$$\hat{\phi}_i = a_i + b_i \qquad (i = 1,2,\ldots,p)$$

$$\hat{\theta}_i = -a_i \qquad (i = 1,2,\ldots,q)$$

5.2 Case Study

The environmental system was identified by using an ADF on SO_2 air-pollution data from February 1 to July 17, 1973, at monitoring station no. 9 in Venice. The following models were considered after estimation of autocorrelation and partial autocorrelation functions (Box and Jenkins, 1976):

$$\text{ARI} \qquad (1 - z^{-24})(1 - \phi_1 z^{-1})x_k = \epsilon_k \qquad \text{(i)}$$

$$\text{ARIMA} \quad (1 - z^{-24})(1 - \phi_1 z^{-1})x_k = (1 + \theta_1 z^{-1})\epsilon_k \qquad \text{(ii)}$$

$$\text{ARIMA} \quad (1 - z^{-24})(1 - \phi_1 z^{-1})x_k = (1 + \theta_1 z^{-1} + \theta_{24} z^{-24})\epsilon_k \qquad \text{(iii)}$$

The results of parameter estimations for these three models using an ADF and 4000 iterations were

$$(\phi_1) = (0.565) \qquad \text{(i)}$$

$$(\phi_1, \theta_1) = (0.383, 0.214) \quad \text{(ii)}$$

$$(\phi_1, \theta_1, \theta_{24}) = (0.431, 0.171, -0.844) \quad \text{(iii)}$$

In the real-data processing it is difficult to detect the order of the time series. To determine the order of the environmental system, Akaike's Information Criterion (AIC) was used (Akaike, 1974). The definition of the AIC for the ARIMA(p,d,q) model is given by Ozaki (1977):

$$\text{AIC}(p,d,q) = N\log(\sigma_\epsilon^2) + \{n/(n-d)\}2(p+q+1+\delta_{do})$$

where σ_ϵ^2 is the variance of the residual and δ denotes Kronecker's delta. The AIC(p,d,q) values corresponding to cases (i)–(iii) described above are found to be

$$\text{AIC}(1,24,0) = 34{,}503 \quad \text{(i)}$$

$$\text{AIC}(1,24,1) = 34{,}634 \quad \text{(ii)}$$

$$\text{AIC}(1,24,24) = 36{,}129 \quad \text{(iii)}$$

Thus the ARI model (i), whose AIC(1,24,0) for the SO_2 data is the minimum found, can be selected as the identification result.

We now describe the performance of the forecasting scheme using an ADF on SO_2 data from May 15 to May 18, 1975, at Komatsushima, Japan, and from June 15 to June 19, 1973, at monitoring station no. 9 in Venice, Italy. The 1-h forecasts are illustrated in Figures 1 and 2 (1000 observations were used to adapt the parameters of the ADF). It can be seen that the dynamics of SO_2 are more rapid in Venice than in Komatsushima. Thus the Venice data can be modeled as ARIMA(3,1,3) or ARIMA(3,24,3). The forecasting result for SO_2 data in Venice is summarized in Table 6.

TABLE 6 The forecasting result for SO_2 data in Venice.

Model	Forecasting error (mean, variance)
Persistence with $(1 - z^{-1})$	(–0.083, 6124)
ADF with $(1 - z^{-1})$	(–0.366, 5566)
Persistence with $(1 - z^{-24})$	(–1.123, 9025)
ADF with $(1 - z^{-24})$	(–0.830, 7038)

The problem of detecting pollution episodes was analyzed by using the error values of the ADF forecasting scheme. The raw data from April 18 to April 21, 1973, at monitoring station no. 9 in Venice contain evidence of a number of pollution episodes. In order to detect these episodes the error of the ADF was used because the ADF "learns" from past pollution levels and does not forecast the rapid changes characteristic of pollution episodes. The detection result (with a threshold check of the raw data and using two ARIMA models) is illustrated in Table 7 as the detection measure J.

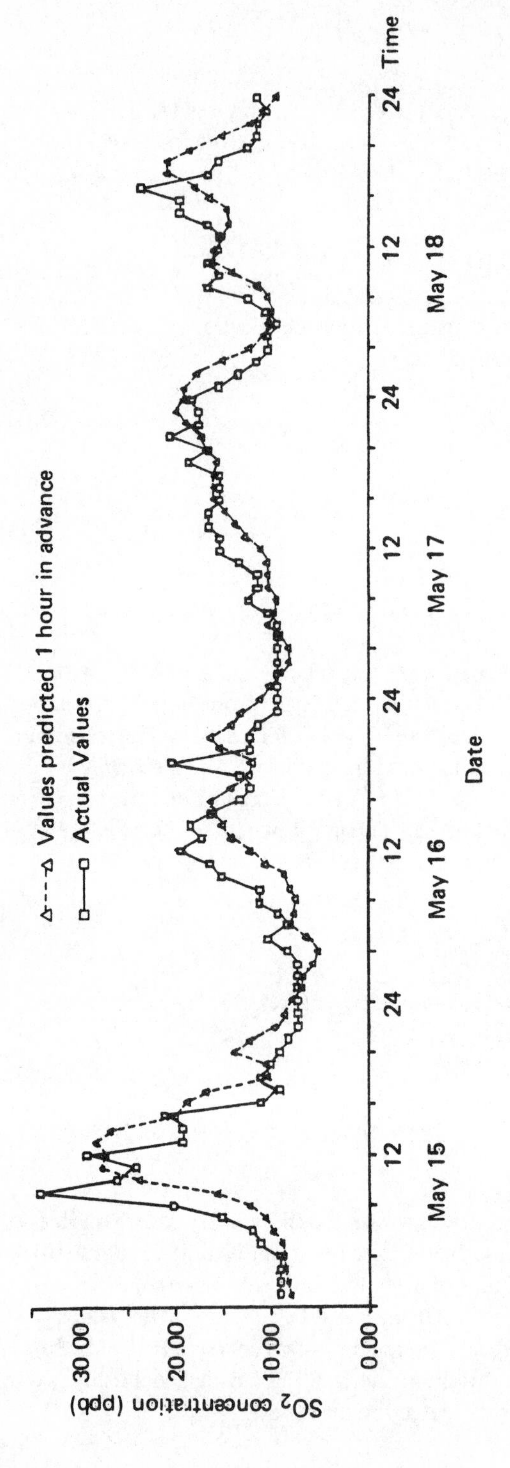

FIGURE 1 Forecasting of SO_2 data at Komatsushima using an ADF.

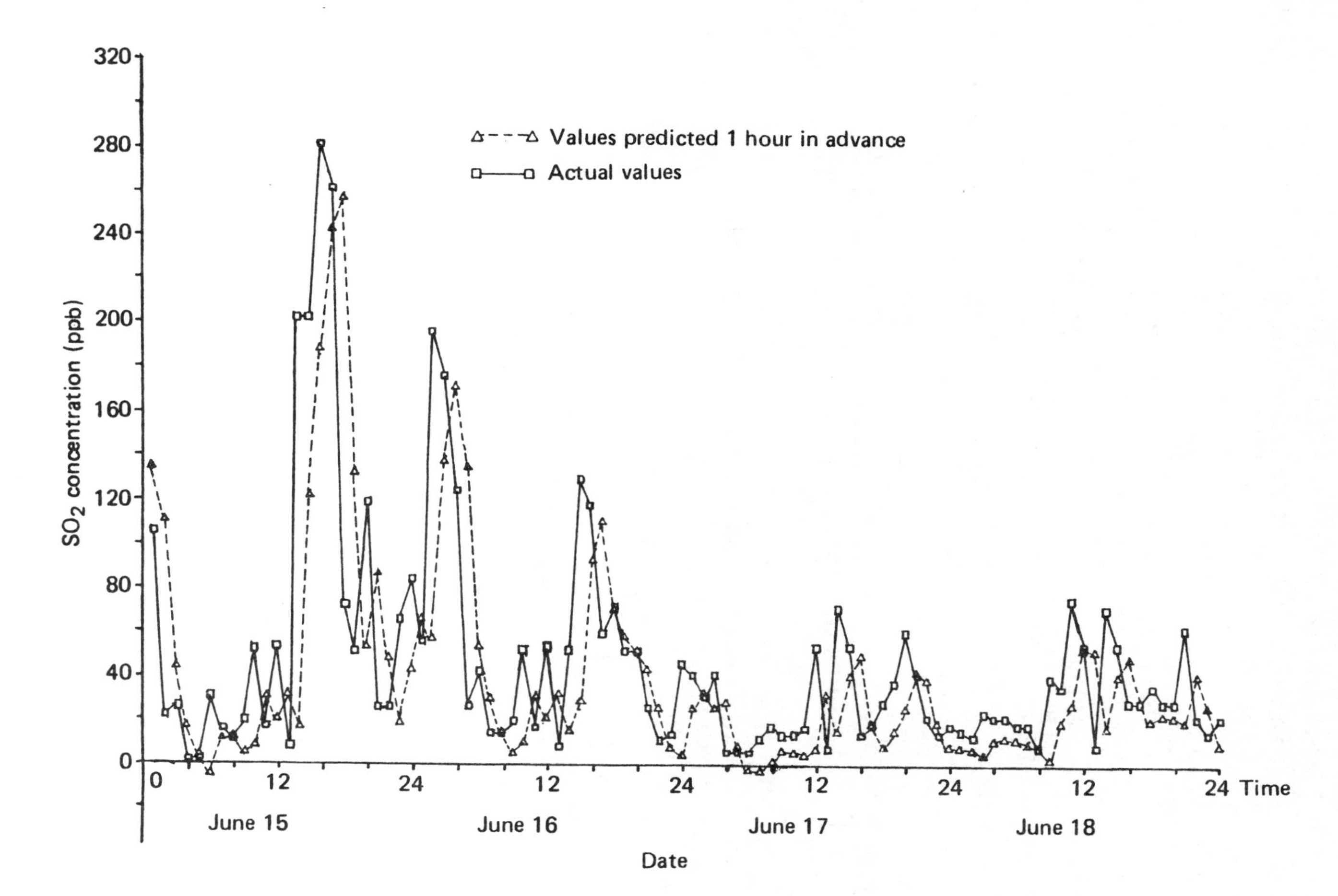

FIGURE 2 Forecasting of SO_2 data in Venice using an ADF.

TABLE 7 The detection measure J using an ADF.

Date in 1973	Threshold check of raw SO_2 data (200 ppb, 250 ppb)	Threshold check of prediction error using the ADF (1.5σ, 2.0σ)		Measure J
		ARIMA(3, 1, 3)	ARIMA(3, 24, 3)	
April 18	(12:00)	(12:00)	12:00	4
	13:00	13:00	13:00	6
	14:00	–	(14:00)	3
	15:00	–	–	2
	16:00	–	–	2
	17:00	17:00	(17:00)	5
	–	–	22:00	2
April 19	9:00	(9:00)	(9:00)	4
	(10:00)	–	–	1
	(14:00)	(14:00)	–	2
	15:00	–	–	2
	18:00	–	18:00	4
April 20	–	–	–	–
April 21	22:00	22:00	22:00	6

The times in parentheses indicate misdetection in the case of the upper threshold.

6 THE KALMAN FILTERING TECHNIQUE FOR FORECASTING POLLUTION LEVELS

The Kalman filtering technique has been applied to the forecasting of pollution levels. The advantages of applying this technique can be summarized as follows. (i) The modeling error is contained in the forecasting equations in the form of covariance matrices. (ii) The concentration data are measured with some degree of error, but the Kalman filter can estimate the pollution levels from the measurements.

Consider the model

$$\boldsymbol{x}(k) = \boldsymbol{A}(k,k-1)\boldsymbol{x}(k) + \boldsymbol{b}(k-1) + \boldsymbol{u}(k-1)$$

$$\boldsymbol{y}(k) = \boldsymbol{x}(k) + \boldsymbol{v}(k)$$

where $\boldsymbol{u}(k)$ and $\boldsymbol{v}(k)$ are n vectors representing independent noises with zero-means and covariance matrices $Q(k)$ and $R(k)$ respectively.

The m-h-ahead optimal forecast is

$$\hat{\boldsymbol{x}}\{k/(k-m)\} = \boldsymbol{A}(k,k-m)\hat{\boldsymbol{x}}\{(k-m)/(k-m)\} + \sum_{j=k-m+1}^{k} \boldsymbol{A}(k,j)\boldsymbol{b}(j-1)$$

where the filtered value $\boldsymbol{x}(i/i)$ is given (Meditch, 1969) by

$$\hat{\boldsymbol{x}}(i/i) = \hat{\boldsymbol{x}}\{i/(i-1)\} + \boldsymbol{K}(i)[\boldsymbol{y}(i) - \hat{\boldsymbol{x}}\{i/(i-1)\}]$$

$$P(i/i) = P\{i/(i-1)\} - K(i)P\{i/(i-1)\}$$

$$K(i) = P\{i/(i-1)\}[P\{i/(i-1)\} + R(i)]^{-1}$$

$$\hat{x}\{i/(i-1)\} = A(i,i-1)\hat{x}\{(i-1)/(i-1)\} + b(i-1)$$

$$P\{i/(i-1)\} = A\{i,(i-1)\}P\{(i-1)/(i-1)\}A^{T}(i,i-1) + Q(i-1)$$

where $P(i/j)$ is the covariance matrix of the estimation error for $x(i/j)$.

Table 8 shows the forecasting error for five pollution levels on December 3, 1971, measured in Tokyo with the performance index

$$J = (1/24) \sum_{k=1}^{24} \| y(k) - \hat{x}\{k/(k-m)\} \| / \| y(k) \|$$

where $\| \ \|$ denotes the euclidean norm and the models were constructed using the measurements from October 1 to November 30. The diagonal elements of $R(k)$ were set to 0.1 and the remaining elements were assumed to be zero. This corresponds to the assumption that the variance of each component of $v(k)$ is slightly larger than the variance of quantization for the measurements (0.0833). From Table 8 it can be seen that the forecasting accuracy is higher for O_x than for the other factors.

TABLE 8 Forecasting accuracy using the Kalman filter.

Kalman filter				Persistence
Measured factor	J	Measured factors	J	J
O_x	0.308	O_x, CO, SO_2	0.331	0.628
CO	0.339	O_x, CO, NO	0.490	
SO_2	0.418	O_x, SO_2, NO_2	0.436	
NO	0.347	O_x, CO, SO_2, NO, NO_2	0.325	
NO_2	0.325			

7 FORECASTING USING THE GROUP METHOD OF DATA HANDLING (GMDH)

7.1 Basic GMDH

The GMDH algorithm (Ivakhnenko, 1970, 1971) can be used to forecast pollution data. The GMDH builds the input–output relationship of a complex system using a multilayered perceptron-type network structure. Each element in the network implements a nonlinear function of its inputs. The function implemented in each element is usually a second-order polynomial with two inputs. In the GMDH algorithm all the experimental data are divided into training and checking data sets. The coefficients in each layer are calculated by using the training data set. Then only the variables satisfying forecasting accuracies for the checking data set are optimized in the next layer.

The practical algorithm is as follows.

(i) Select L elements which have a strong correlation with the true values to be forecast.
(ii) (For training data) Calculate the least-square estimates of the polynomial coefficients by solving the normal equation with the outputs from the preceding layer as inputs to this layer. The number of outputs in this layer is the number of pairwise combinations of L variables.
(iii) (For checking data) Select L variables minimizing the variance of the forecasting error from the outputs in this layer.
(iv) If the variance is greater than or equal to that in the preceding layer, then the modeling is stopped and the forecasting is implemented. If the variance is less than that in the preceding layer, go to step (v).
(v) Proceed to the next layer and continue the nonlinear modeling. (Go to step (ii).)

7.2 Revised GMDH

In the use of the basic GMDH it is necessary to divide the measurements into training and checking data sets. In order to avoid this heuristic division we consider a revised GMDH (Tamura and Kondo, 1977; Sawaragi et al., 1979). The significant advantage of the revised GMDH algorithm is that it does not necessitate the division of the available data into two data sets. All the data can be used for constructing the nonlinear model and at the same time for evaluating the forecasting error. Instead of the Prediction Sum of Squares (PSS) method (Tamura and Kondo, 1977), the AIC is used here for evaluating the performance of parameter estimates. The revised GMDH is based on the following procedures.

(i) (Generation of the optimal polynomials in each selection layer) The optimal partial polynomials can be generated by applying a stepwise regression procedure for the input variables to the second-order polynomial with one or two variables. In this procedure, AIC is used as a criterion for selecting dominant variables. For multiple regression analysis AIC is defined as

$$\mathrm{AIC} = N \log(\hat{\sigma}_n^2) + 2(n+1) + C$$

where $\hat{\sigma}_n^2$ is the variance of the forecasting error, n the number of independent variables, N the number of data used, and C a constant.
(ii) (Selection of the intermediate variables) The L intermediate variables which give the L smallest AIC values are selected from all the intermediate variables.
(iii) (Stopping of the multilayered interactive computation) When all the generators of the optimal partial polynomials in the selection layer become a first-order polynomial with a single variable, the iterative computations of the revised GMDH are terminated because the values of the AIC cannot be decreased any

further. The forecasting model is obtained as a weighted average of the complete polynomials which are constructed by the intermediate variables remaining in the final layer.

7.3 Case Study

The forecasts of SO_2 data measured in Tokushima and Venice by the GMDH are shown. In particular, the following results are discussed: (i) A comparison between ARIMA forecasts and GMDH forecasts; (ii) A comparison between the forecasts for rapidly-varying data and slowly-varying data; (iii) A comparison between the Venetian and Japanese data.

7.3.1 Forecasting of Tokushima SO_2 Data

(a) The monitoring stations and their main characteristics were as follows. Monitoring stations Kawaguchi and Matsushige are located near the airport and factories; the prevailing wind is strong, and the data undergo rapid variation. Monitoring stations Kitagima and Aizumi, on the other hand, are characterized by a relatively slow variation in data.

(b) The input data for the basic GMDH were as follows: (i) SO_2 pollution levels measured hourly; (ii) wind velocity, wind direction (southeasterly component), and the mean value of the intensity of the wind velocity, measured at the four monitoring stations; (iii) weather conditions (classified as fine, cloudy, or rainy). The missing data were calculated by interpolation.

(c) The training and checking data were as follows. SO_2 and meteorological data from May 1 to May 10, 1975, were used; eight days with large variance were selected from the ten days as training data, and two days with small variance were used as checking data.

(d) Forecasting was for SO_2 pollution levels from May 11 to May 12, 1975. The results are illustrated in Table 9 and Figure 3.

TABLE 9 Comparison of forecasts using the GMDH.

	Variance of forecasting error	
Monitoring station	1 h in advance	3 h in advance
Kawaguchi	2.45	3.00
Matsushige	3.94	5.65
Kitagima	2.08	2.93
Aizumi	1.57	1.66

7.3.2 Forecasting of Venice SO_2 Data

(a) The monitoring station was the no. 9 station in Venice (at one of the most polluted sites) and the main characteristic was a rapid variation in data. It should be noted that pollution episodes occur.

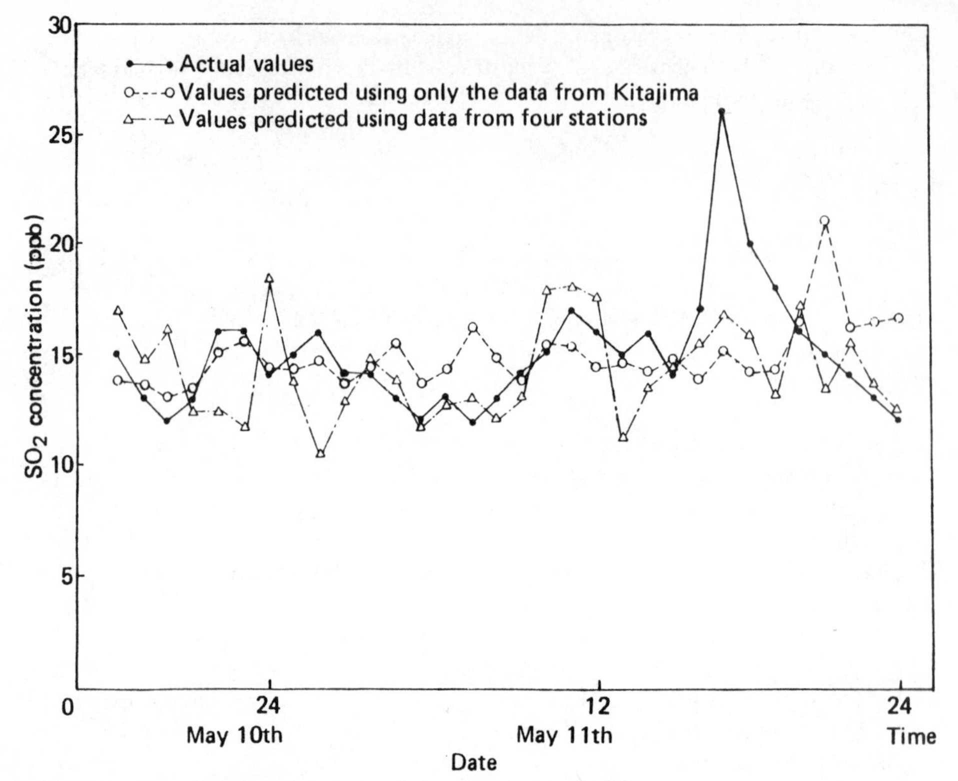

FIGURE 3 Forecasting of SO_2 data at Kawaguchi using the GMDH.

(b) The input data for the basic GMDH were as follows: (i) SO_2 pollution data measured hourly at monitoring stations nos. 2, 6, 9, and 10; (ii) Wind velocity and wind direction (measured in eight 45° sectors of the compass).

(c) The training data were from May 26 to June 2, 1973, and the checking data were from June 3 to June 4, 1973.

(d) Forecasting was for SO_2 data at monitoring station no. 9 from June 5 to June 8, 1973. The result is illustrated in Figure 4.

(e) By comparing the result in Figure 4 with the ADF forecasting of the ARIMA model shown in Figure 2 it can be clearly seen that the GMDH forecasts have no time lag and give forecasts higher than the daily maximum. Therefore in order to forecast pollution episodes in Venice it is better to use the GMDH algorithm.

7.3.3 Forecasting using the Revised GMDH

(a) The monitoring station was Tokushima and the main characteristic was a slow variation of data.

(b) The input data for the revised GMDH are SO_2 pollution data, wind velocity, and wind direction.

(c) Figure 5 shows the advantage in 3-h forecasting accuracy using the revised GMDH model compared with the time-series models. However, the revised GMDH needs much more computational time for model building than do the linear time-series models.

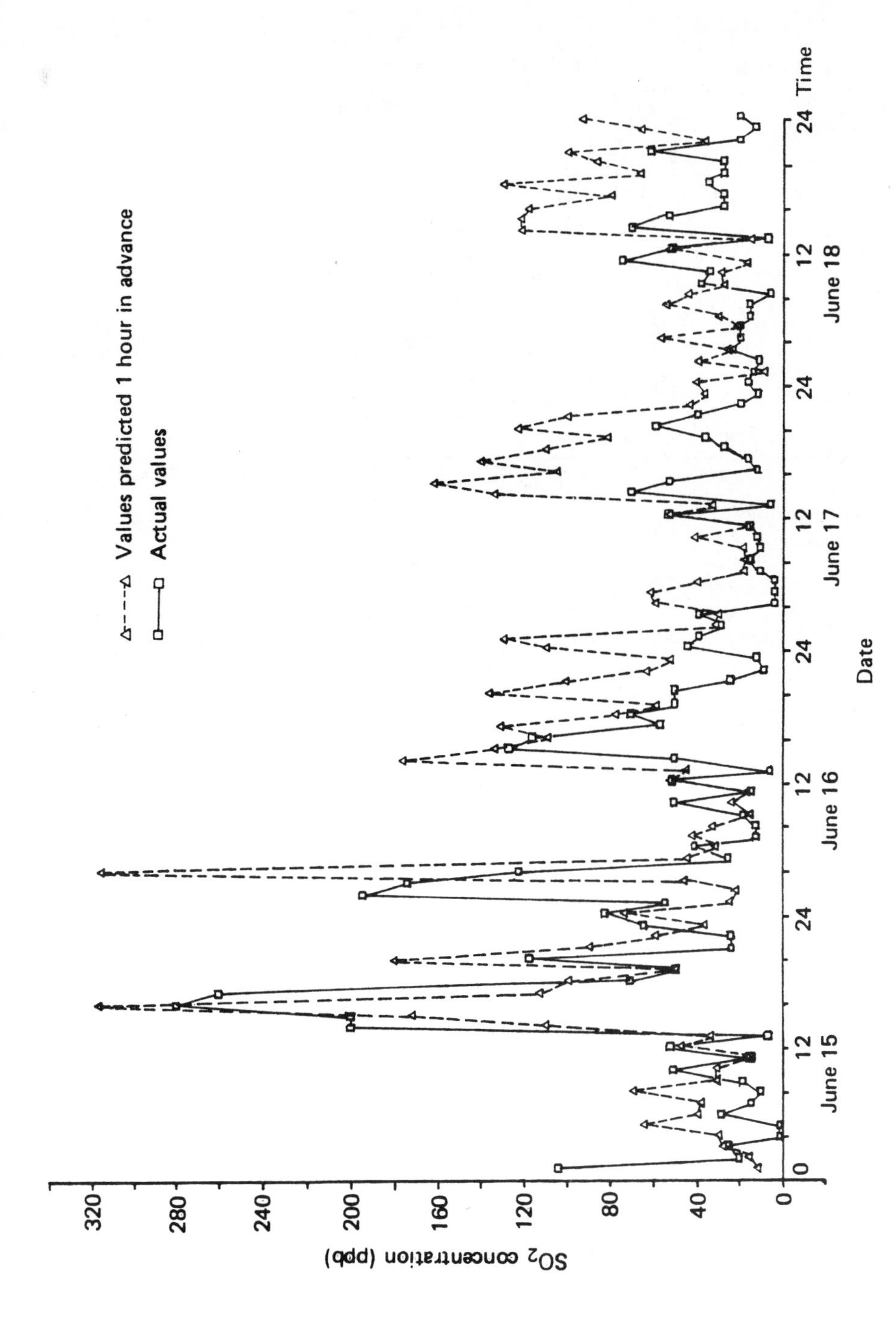

FIGURE 4 Forecasting of SO_2 data in Venice using the GMDH.

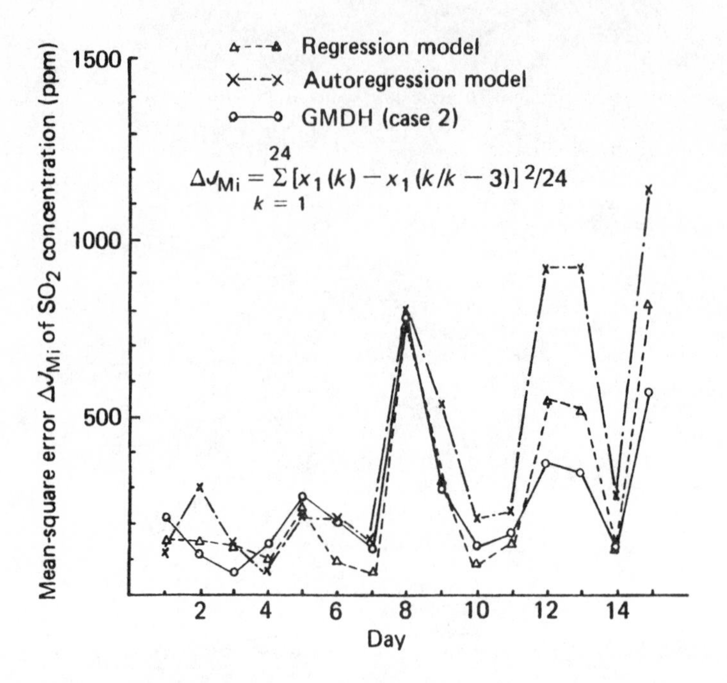

FIGURE 5 Comparison of the forecasting accuracy of three models.

8 CONCLUSIONS

The accuracy of forecasts of air-pollution levels by the multiple linear regression, the AR, the Box–Jenkins, and the persistence models have been discussed. In addition, forecasting methods based on an ADF, the Kalman filtering technique, and the GMDH have been illustrated. For the four kinds of performance index, the multiple linear-regression model resulted in a better forecasting accuracy than other time-series models, particularly in the forecasts 2 h or 3 h in advance. The ADF has a simple structure and is suitable for detecting pollution episodes. The Kalman filtering technique modifies the estimates of pollution levels by the innovation sequence. The GMDH is better than the multiple linear-regression model but needs much more computational time. The advantage of the GMDH is its ability to forecast pollution episodes.

When the forecasts of pollution levels are higher than the safety level, the pollution sources must be controlled. However, it is very difficult to decide how to control them because non-physical models do not examine the role of emission. A future study will be devoted to the derivation of the relationship between non-physical and physical models and to the design of a model which combines control and forecasting. Furthermore, investigations of the feasibility of such models in operative decision making will be carried out.

REFERENCES

Akaike, H. (1969). Fitting auto-regressive models for prediction. Ann. Inst. Math. Stat., 21:243.

Akaike, H. (1970). Statistical prediction identification. Ann. Inst. Math. Stat., 22:203.

Akaike, H. (1974). A new look at statistical model identification. IEEE Trans. Autom. Control, 19:716.

Akizuki, J. and Shirai, K. (1975). On construction of air pollution models by a statistical method. In Proc. Symp. Modeling for Prediction and Control of Air Pollution. p. 53.

Bendat, J.S. and Piersol, A.F. (1971). Random Data: Analysis and Measurement Procedures. Wiley, New York.

Box, G.E. and Jenkins, G.M. (1976). Time Series Analysis: Forecasting and Control. Holden–Day, San Francisco, California.

Desalu, A.A., Gould, L.A., and Schweppe, F.C. (1974). Dynamic estimation of air pollution levels. IEEE Trans. Autom. Control, 19:904.

Ishihara, H. and Soeda, T. (1976). On the determination of the optimal observation time points for the reconstruction of the linear state subjected to random inputs. Trans. SICE, 12:162.

Ivakhnenko, A.G. (1970). Heuristic self-organization in problems of engineering cybernetics. Automatika, 6:207.

Ivakhnenko, A.G. (1971). Polynomial theory of complex systems. IEEE Trans. Syst., Man, Cybern., 1:364.

Kikuchi, A., Omatu, S., and Soeda, T. (1979). Applications of adaptive digital filtering to the data processing for an environmental system. IEEE Trans. Acoust., Speech, Signal Process., 27:790.

Kondo, J. (1975). Air Pollution. Koronasha, Tokyo.

McCollister, G.M. and Wilson, K.R. (1975). Linear stochastic model for forecasting daily maxima and hourly concentrations of air pollutants. Atmos. Environ., 9:417.

Meditch, J.S. (1969). Stochastic Optimal Linear Estimation and Control. McGraw-Hill, New York.

Ozaki, T. (1977). On the order determination of ARIMA models. Appl. Statist., 26:290.

Sawaragi, Y., Soeda, T., Yoshimura, T., Ohe, S., Chujo, Y., and Ishihara, H. (1976). The prediction of air pollution levels by non-physical models based on the Kalman filtering method. Trans. ASME, 98-G:375.

Sawaragi, Y., Soeda, T., Tamura, T., Yoshimura, T., Ohe, S., Chujo, Y., and Ishihara, H. (1979). Statistical prediction of air pollution levels using non-physical models. Automatika, 15:441.

Soeda, T. and Ishihara, H. (1974). Application of Kalman filtering method and *n*-observability principle to the prediction of the atmospheric pollution levels. In Proc. Symp. Nonlinear Estimation and its Applications, 5th, San Diego, California.

Tamura, H. and Kondo, T. (1977). Large-spatial pattern identification of air pollution by a combination of source–receptor matrix and revised GMDH. In Proc. IFAC Symp. Environmental System Planning, Design and Control. p. 378.

White, S.A. (1975). An adaptive recursive digital filter. In Proc. Asilomar Conf. Circuits, Systems and Computers, 9th. p. 21.

Widrow, B., et al. (1975). Adaptive noise cancelling: principles and applications. Proc. IEEE, 63:1692.

REAL-TIME CONTROL OF EMISSIONS IN JAPANESE CITIES

T. Soeda and S. Omatu
Tokushima University, Tokushima (Japan)

1 INTRODUCTION

In recent years, the Japanese economy has undergone great development, and the GNP has grown to rank second in the world – only that of the USA is now larger. Indeed, the Japanese people have attained the highest degree of affluence in their history. However, the rapidly growing economy has caused serious social difficulties and environmental deterioration, in particular, has become a nationwide problem.

Standards have been established for air and water pollution control, and efforts have been made to clean up the environment, with some positive results. However, air-pollution levels (mainly from SO_2 and NO_x pollutants) still vary widely, due to the broad range of atmospheric conditions. In view of this variability, concentrations can still occasionally become intolerably high, even though "proper" emissions levels have been set beforehand. For this reason, Japanese air-pollution legislation considers the possibility of implementing real-time emission control as an emergency measure. For instance, when the SO_2 concentration exceeds 0.5 ppm for three consecutive hours, such emergency measures are taken by the local governments.

Of course, if concentrations of SO_2 and NO_x could be accurately forecast and emission sources controlled accordingly, even these occasional incidents of high pollutant concentration could be avoided. The Air Pollution Monitoring System (APMS) allows the forecasting of 3-h ahead concentrations and subsequently the control of emissions, so that the air-quality standards mentioned in Table 1 can be met.

Since 1971, development of the APMS has been undertaken by a research group (with about 30 members) organized by the Japanese Society for the Promotion of Machine Industry. Forecasting techniques discussed by the research group have been summarized by Yokoyama (1973). Parallel to these discussions, APMS hardware was built at the Kashima industrial complex in July 1975, and now the entire system is being tested.

Early in 1972, the environmental control group of the Tokyo Scientific Center, IBM Japan, decided to develop computer technology related to air-pollution problems. Meanwhile, in August 1972, the Hyogo Prefecture organized the Air Pollution Prediction Special Research Team, with the purpose of developing air-pollution regulations, industrial zoning, and urban planning. The environmental control group set up a computer

TABLE 1 Ambient air-quality standards in Japan.

Pollutant	Standard
Sulfur oxides	1. Hourly average concentration must be below 0.1 ppm for more than 99% of the hours in a year. 2. Daily average concentration must be below 0.04 ppm for more than 98% of the days in a year.
Nitrogen oxides	1. Daily average concentration must be below 0.02 ppm for more than 98% of the days in a year.

model and checked its effectiveness by using data and related information supplied by the Hyogo team. The urban area tested was Himeji City, in Hyogo Prefecture.

This paper describes current pollution control in Japanese cities and some of the case studies mentioned above. Experience in real-time emission control in the industrial area of Kashima is illustrated in Section 2. Section 3 describes current real-time control measures in Japanese urban centers, as well as control trends in Osaka and Tokyo. The Himeji urban case study mentioned above is discussed in detail in Section 4.

2 REAL-TIME CONTROL OF POLLUTION IN AN INDUSTRIAL AREA: KASHIMA CASE STUDY

The real-time control case study which took place in Kashima industrial complex and which was developed by J. Sakagami, M. Hino, and O. Yokoyama, will now be discussed.

2.1 Data Collection in Kashima Industrial Area

Kashima district is a newly built industrial complex located on the coast of the Pacific Ocean, ca. 100 km from Tokyo. As shown in Figure 1, the industrial area extends ca. 20 km along the coast and 7 km inland. There are steel mills (area No. 1 on the map), petro-chemical industries (No. 2), and power stations (No. 3). The total amount of SO_2 and NO_x emitted from these sources is ca. 5000 and 2000 Nm/h, respectively, while the level of automobile emission of NO_x in the area is about 200 Nm/h.

Emissions of SO_2 and NO_x are monitored by 44 stations. Data recorded are hourly average concentrations of SO_2 and NO_x, speed and temperature of effluent gas, rate of consumption of oil, power-generation rate, and air-blower power.

The local meteorological factors which are most relevant for diffusion processes are surface wind velocity, intensity of turbulence, insolation and reverse infrared radiation (net radiation flux), vertical temperature profile, and rate of rainfall. Ground-level wind velocity is measured at the same stations where concentration measurements are taken. Temperature and wind profiles and intensity of turbulence are measured at heights of 50, 110, and 220 m, by sensors located on a high stack (the black square in Figure 1). Net radiation flux and rate of rainfall are measured at one station. The height of the "lid"

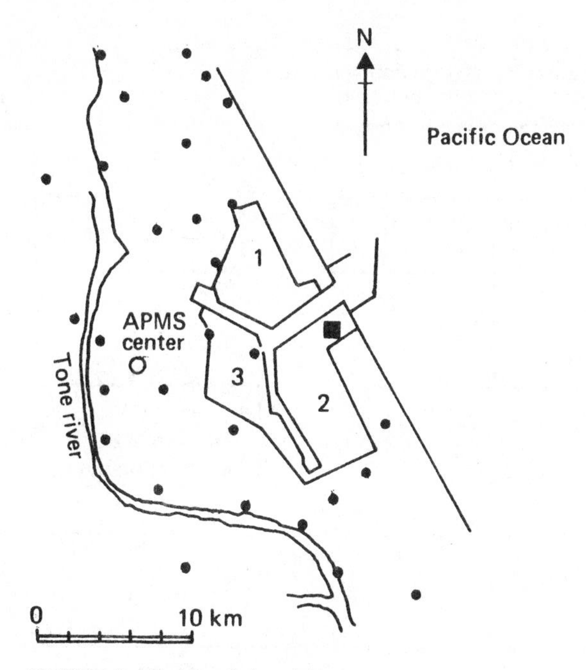

FIGURE 1 Kashima industrial area.

created by temperature inversion is measured by using a remote sensing temperature profiler (PITS). Furthermore, synoptic weather data, recorded by the meteorological agency, are communicated every 3 h through the Japanese Meteorological Society. All data are sent to the Kashima APMA center (shown by the circle in the map) by telephone line every hour, except for upper wind data which are sent every 10 minutes.

The data sent by telephone line are recorded and processed by 2 minicomputers (NEAC-M4) and then sent to a central computer (NEAC-2200-500) for the forecasting calculations.

2.2 Forecast and Control of Pollution

The procedure for forecasting and controlling emissions is illustrated in Figure 2. First, the possibility of high concentrations occurring during the next day is investigated by using discriminatory analysis. When it is decided that low concentrations are most probable, the following steps are unnecessary.

If high pollutant concentrations seem probable, the forecast of those meteorological variables which are relevant for the diffusion processes is carried out by statistical methods. SO_2 and NO_x emissions are also forecast by statistical methods. Future environmental concentrations of the two pollutants are estimated both by statistical techniques and by a diffusion model; the latter also assesses the contribution of each polluter. If the forecast concentrations exceed an assigned level (predetermined on the basis of the

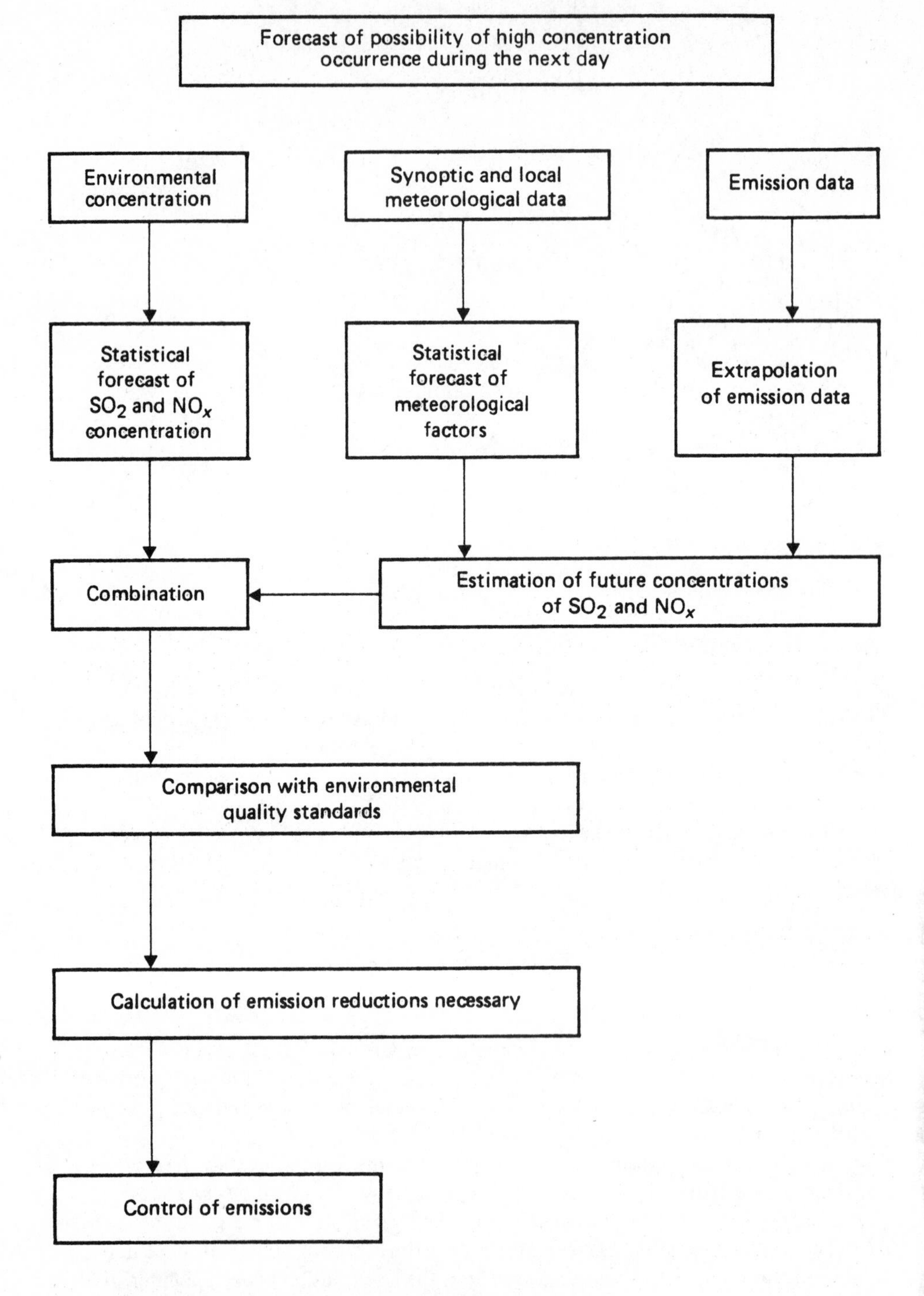

FIGURE 2 Real-time emission-control scheme for Kashima industrial area.

ambient air-quality standards given in Table 1), the emission of each polluter is reduced according to its contribution to the overall environmental concentration.

The individual steps of the procedure will now be illustrated in detail.

2.2.1 Discriminatory Analysis on the Occurrence of High Pollutant Concentrations

Any day during which a particular hourly concentration of a pollutant exceeds a given level, C_c, at some monitoring stations, is considered to be a "polluted day" and is denoted by the index w. Conversely, $\overline{w}$ denotes "nonpolluted days". $x_1, x_2, \ldots, x_p$ are the factors used in the discriminatory analysis, which is based on the index

$$I_p = \sum_{i=1}^{p} a_i x_i \qquad (i = 1,2,\ldots,p) \tag{1}$$

Pressure, wind velocity, thermal stability, and rainfall intensity at various sites in Japan have been taken as forecasting factors. For example, the pressure factor x_1 is the pressure difference between Wajima and Choshi meteorological stations. The coefficients a_i in eqn. (1) are determined by one of the following criteria:

(a) Maximize $|\overline{I_p(\mathrm{w}) - I_p(\overline{\mathrm{w}})}|$;

(b) Minimize the variance of $I_p(\mathrm{w})$ and $I_p(\overline{\mathrm{w}})$;

(c) Maximize $R = |\overline{I_p(\mathrm{w})} - \overline{I_p(\overline{\mathrm{w}})}|/(S^2(\mathrm{w}) + S^2(\overline{\mathrm{w}}))$

where $I_p(\mathrm{w})$ and $I_p(\overline{\mathrm{w}})$ denote I_p on a polluted day and on a nonpolluted day, respectively, $S(\mathrm{w})$ is the standard deviation of $I_p(\mathrm{w})$, and the bar denotes time average. The threshold concentration C_c is given by

$$C_c = (N_1 \overline{I_p(\mathrm{w})} + N_2 \overline{I_p(\overline{\mathrm{w}})})/(N_1 + N_2)$$

where N_1 and N_2 are the numbers of polluted and nonpolluted days, respectively. An example of the forecast is shown in Table 2.

TABLE 2 Discriminatory analysis on the occurrence of a "polluted day".

	Observed	
Forecast	Polluted days	Nonpolluted days
Polluted days	1300	3100
Nonpolluted days	132	2395

2.2.2 Statistical Forecast of Meteorological Factors

Meteorological factors such as wind velocity, thermal stability, and turbulence intensity are forecast 1- and 3-h ahead by using canonical correlation analysis. As an example of this procedure, the wind-velocity forecast is described briefly.

Wind velocity is first divided into its northern and eastern components and each component is subdivided into mean and departure from the mean. For each component, the departure $U(k,t)$ (k is the component index; t is the time index) is forecast. Specifically, according to principal component analysis, it is first assumed that

$$U(k,t) = \sum_{i=1}^{N} Y_i(t) s_i(k) \tag{2}$$

where $s_i(k)$ are orthogonal vectors, $Y_i(t)$ are coefficients, and N is the number of components. The coefficients $Y_i(t)$ are predicted by the following extrapolation formula:

$$Y_i(t) = a + bt \tag{3}$$

where a and b are determined through least mean-square error estimation, applied to the data of the last 24 h. A sample of forecast performance is given in Figure 3.

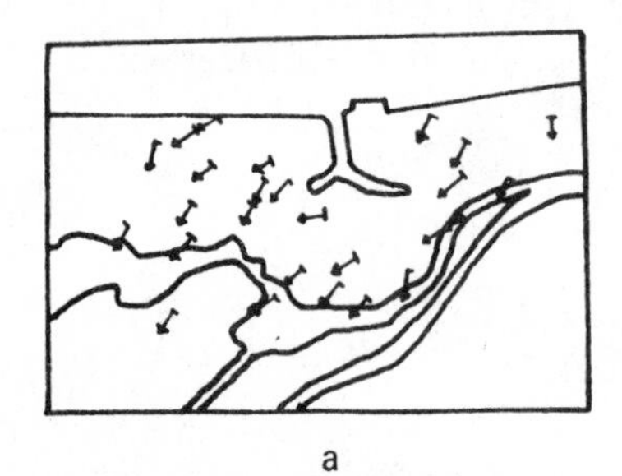

a

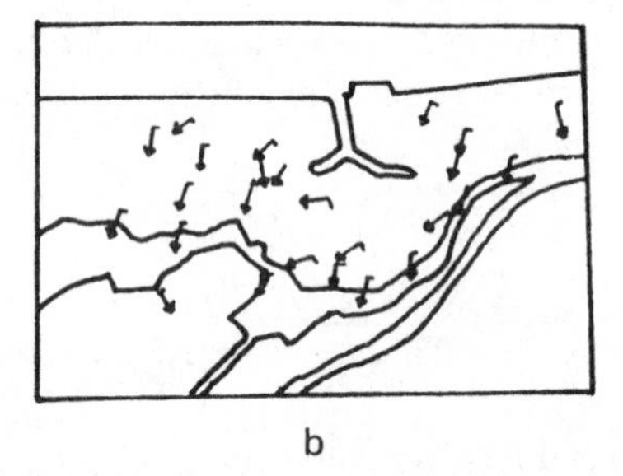

b

FIGURE 3 (a) Measured and (b) forecast wind velocity at 2 p.m. on 27 August 1975.

2.2.3 Prediction of Emission

The emission data used by the diffusion model described below (see Section 2.2.4) are stack height and diameter, temperature and flow-rate of effluent gas, and gas concentration of SO_2 and NO_x. In particular, gas flow-rate is estimated by measures of gas speed, and rate of fuel or electrical-power consumption (by using empirical relationships). The amount of effluent gas, temperature, and concentration is predicted 3-h ahead by extrapolating present and past measures.

2.2.4 Forecast of SO_2 and NO_x Concentrations by a Statistical Method

The concentrations of SO_2 and NO_x are forecast 1- and 3-h ahead by a method which combines ordinary multiple regression with Kalman filtering (Hino, 1974). First the number of forecasting factors is reduced by principal component analysis (see the method described in Section 2.2.2). Then, each forecasting factor f_m is subdivided into a periodic component f_p and a random component f:

$$f_m = f_p + f$$

The random component is forecast by statistical methods. More precisely, f is first predicted by using multiple regression analysis and then the prediction error is reduced by using the Kalman filter. Thus, each factor at time $(i + 1)$ is forecast by the following relationship, which takes into account the latest measurements:

$$\mathbf{F}_p(i+1) = \sum_{j=1}^{N} \mathbf{A}(j-1)\,\mathbf{F}(i-j-1) \tag{4}$$

where **F** is the kth vector of f, **A** is a $k \times k$ matrix, and p denotes prediction. The coefficient **A** can be determined by self and cross correlations of f-data, according to Wiener's criterion (minimization of r.m.s. prediction error).

The forecast by Kalman filter is formally written as:

$$\begin{aligned} \mathbf{F}_p(i+1) &= \mathbf{A}(i+1/i)\,\mathbf{F}_p(i) + K(i)\,\mathbf{F}(i) \\ \Delta\mathbf{F}(i) &= \mathbf{F}(i) - M(i)\,\mathbf{F}_n(i) \end{aligned} \tag{5}$$

where $\mathbf{A}(i + 1/i)$ is the transition matrix, $K(i)$ is the optimal filter gain, $\mathbf{F}(i)$ is the prediction error, and $M(i)$ is the ratio between $\mathbf{F}_p$ and $\mathbf{F}$.

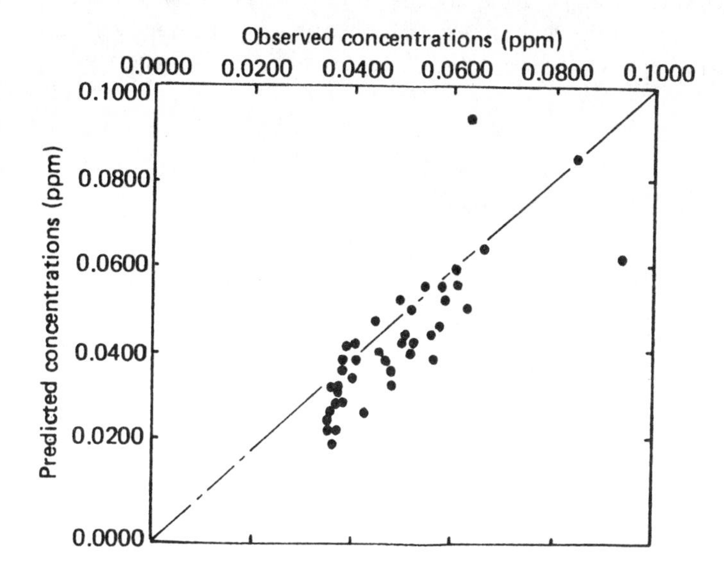

FIGURE 4 Observed values vs. 3-h ahead forecasts for SO_2 concentrations on 2 and 3 August 1975.

Some samples of 3-h SO_2 forecast are shown in Figure 4. The forecast performance by multiple regression analysis and by the present "combined" method are summarized for comparison in Table 3.

2.2.5 Estimation of SO_2 and NO_x Concentrations by a Diffusion Model

A diffusion model has no predictive capability by itself, but estimates of future pollution concentrations and of the contribution of each source can be obtained by introducing

TABLE 3 Correlation between forecasts and observations by the two prediction methods.

	Combined method		Regression	
Station No.	1 h	3 h	1 h	3 h
101	0.88	0.75	−0.099	−0.11
102	0.80	0.77	0.73	0.02
104	−0.06	0.23	0.06	0.10
120	0.86	0.77	0.85	0.52

meteorological and emission forecasts into the model. According to APMS methodology, the normal diffusion model has been modified so that it can be applied to nonstationary meteorological conditions. The basic formula for computing the concentrations is

$$C(x,y,z) = qD(y)D(z)/2\pi u \sigma_y \sigma_z \tag{6}$$

where $C(x,y,z)$ is concentration at point (x,y,z), caused by an emission source at height H_e located at the origin (x is direction downwind, z is the vertical coordinate); q is the source strength; u is mean wind speed; σ_y and σ_z are the "widths" of the plume – more precisely the standard deviation of the concentration distribution along the y and z axes, respectively; $D(y)$ and $D(z)$ represent concentration distributions along the y and z axes in the following sense:

$$D(y) = \exp(-y^2/2\sigma_y^2)$$

$$D(z) = \sum_{n=-\infty}^{+\infty} \{\exp[-(z - H_e + 2nL)^2/2\sigma_z^2] + \exp[-(z + H_e + 2nL)^2/2\sigma_z^2]\}$$

where L is height of the lid and n is the number of reflections of smoke between the lid and the ground.

The effective stack height, H_e, is the sum of the stack height and the height of smoke ascent caused by buoyancy and speed of smoke effluent, as estimated using the empirical relationships of Moses and Carson (APMS, 1973). The plume widths can be estimated from the intensity of turbulence or from the surface wind and the net radiative flux (to be measured and forecast). Details of such methods have been given in a preliminary report of the APMS (1975).

Some sample estimates of SO_2 concentration are shown in Figure 5 (here the estimates do not include background concentration). Correlations between the estimate and the measured concentration at various stations are given in Table 4 (0.7 is obtained as the average correlation).

3 REAL-TIME CONTROL OF AIR POLLUTION IN URBAN AREAS

3.1 General Characteristics of Real-Time Control Policy

Various municipalities in Japan try to control air pollution by reducing emissions in real-time. For example, Osaka, the most advanced city from this viewpoint, has set up the following very simple real-time control scheme.

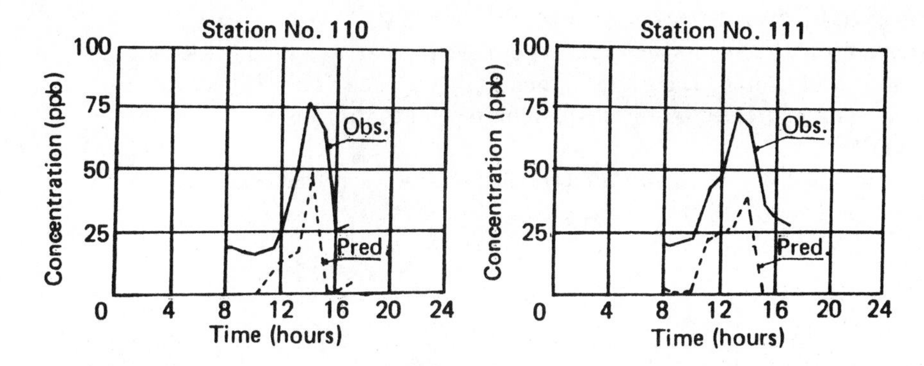

FIGURE 5 Predicted vs. observed (diffusion model) SO_2 concentrations in two monitoring stations on 10 October 1975.

TABLE 4 Correlation between predictions and observations (diffusion model).

Station No.	Correlation
106	0.71
107	0.50
108	0.90
109	0.94
110	0.86
111	0.67
112	0.89
113	0.75
114	0.55

If the measured pollutant concentration exceeds a predetermined level, one of three types of announcement is made:

(a) smog information, when the average SO_2 concentration remains above 0.2 ppm during three consecutive hours;
(b) smog warning, when the average SO_2 concentration remains above 0.5 ppm during three consecutive hours;
(c) smog alarm (implying intervention), when the average SO_2 concentration remains above 0.5 ppm during three consecutive hours.

Of course, the expected weather conditions are empirically taken into account when an announcement is delivered. Almost all Japanese cities now apply the same procedure, which does not involve the use of any complex mathematical model or scheme derived from modern control theory.

In practice, there are in fact some difficulties in implementing the control schemes proposed by modern control theory – the system is "large", i.e. it is characterized by a high number of variables and modeling is complex, due to the complexity of the physical process itself. However, the use of modern control theory still seems feasible. For instance, a control based on real-time predictions seems promising. More specifically, due to the uncertainties in modeling an environmental system, an adaptive predictive control seems recommendable. It seems likely that this type of control will be implemented in Osaka in the future.

Current case studies concerning real-time prediction and "modern" emission control in Osaka and Tokyo Prefectures are illustrated.

3.2 Osaka Case Study

In recent years, air pollution in Osaka City, especially that caused by sulfur oxides from heavy oil burning, has increased rapidly. Thus, the number of days characterized by a "smog information" announcement has increased year after year. In order to avoid excessive pollution of the environment, Osaka Prefecture set up the Public Harm Supervising Center (PHSC) in 1968. The overall organization for controlling air pollution has been established in this center and 15 monitoring stations have been built (Figure 6). Pollution and weather data are transmitted continuously from the monitoring stations to PHSC and the center processes the data immediately. The items monitored by Osaka Prefecture are given in Table 5. The control scheme presently under consideration by Osaka Prefecture is shown in Figure 7.

The first step is the setting up of a mathematical model able to describe pollution dispersion in time and space. Usually, the steady-state solution of the advection–diffusion equation is the model proposed by air-pollution researchers. However, in the Osaka urban air pollution case, it is better to consider a different mathematical model, because of the complexity of the terrain and the nonstationary weather conditions. Thus, Osaka Prefecture has proposed the following mathematical model, which is a modification of the steady-state solution taking account of the city's particular characteristics

$$C_{Aj} = \sum_{\substack{\text{radial} \\ \text{direction}}} [(\Delta\tau_j/\tau)_n \bar{Q}_A / \bar{p}\bar{u}X_E^2\theta]$$

$$C_{pj} = [(\Delta\tau_j/\tau)_n \bar{Q}_p / \sqrt{2\pi}\bar{p}\bar{q}\bar{u}X^2] \exp(-H/\bar{p}X)$$

$$C_B = \sum_j (C_{Aj} + C_{pj})$$

$$dC/dt = \sum_{\text{in}} (Q/V) + \gamma(C_B - C) - \phi$$

where

C = concentration in the "virtual" zone around the monitoring station;
C_B = concentration above the "virtual" zone;
C_A = concentration due to the area emission sources;
C_p = concentration due to the point emission sources;
$\bar{Q}_A$ = intensity of the area emission sources;
$\bar{Q}_p$ = intensity of the point emission sources;

$\bar{u}$ = wind velocity;
H = height of the point emission sources;
$\bar{p}, \bar{q}$ = diffusion parameters in the vertical and horizontal directions, respectively;
$\Delta \tau_j / \tau$ = unit interval in the jth direction per unit time;
ϕ = decreasing concentration rate, because of rainfall or chemical reaction;
γ = diffusion parameter above the zone;
X = distance from the point source to the monitoring station;
X_E = equivalent distance from an area source to the monitoring station, when an area source is regarded as a point source;
Q = emission intensity into the virtual zone;
V = volume of the virtual zone;
n = wind direction index

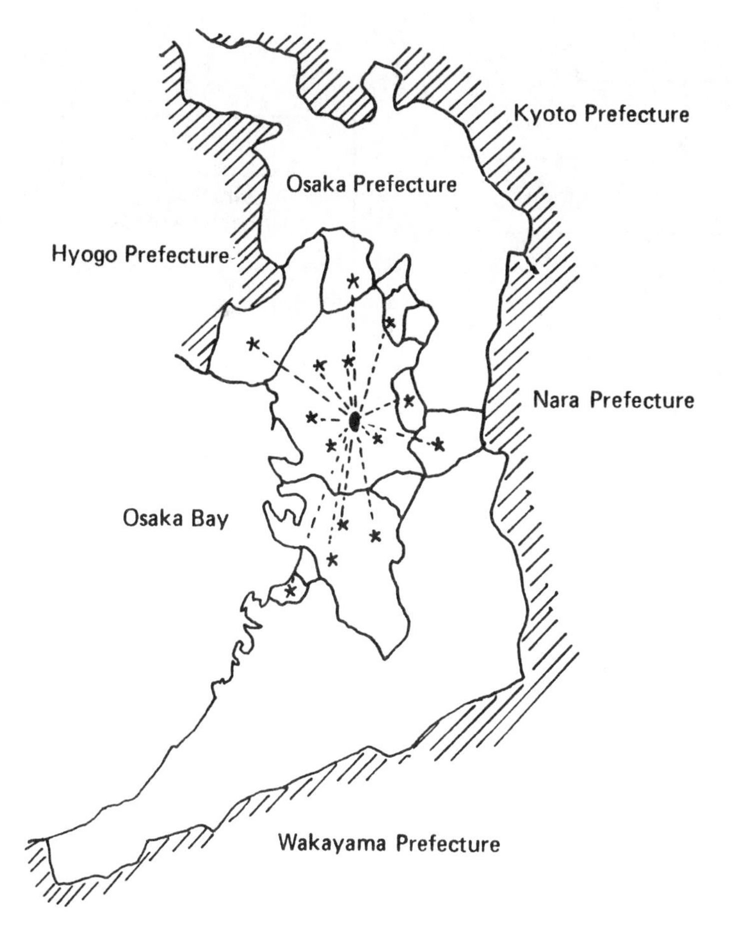

FIGURE 6 Monitoring network in Osaka Prefecture (• = PHSC, * = monitoring station).

TABLE 5 Pollution and weather data monitored in Osaka Prefecture.

Variables monitored	Observed ranges	Types of sensor
Sulfur oxide gas (SO_2 + SO_3)	0–0.2, 0–0.5, 0–1.0 (ppm)	Stationary and mobile
Floating dust	0–1, 0–5, 0–10 (ppm)	Stationary and mobile
Low wind velocity	0.4–10 (m/s), average for 10 min	Stationary and mobile
Wind direction	16 directions, average for 10 min	Stationary and mobile
Temperature	−10 to +50°C	Stationary and mobile
Humidity	0–99%	Stationary and mobile
NO_2	0–0.5, 0–1.0 (ppm)	Mobile
NO	0–0.5, 0–1.0 (ppm)	Mobile
CO	0–50, 0–100 (ppm)	Mobile
C_mH_n	0–100 (ppm)	Mobile
O_3	0–100	Mobile

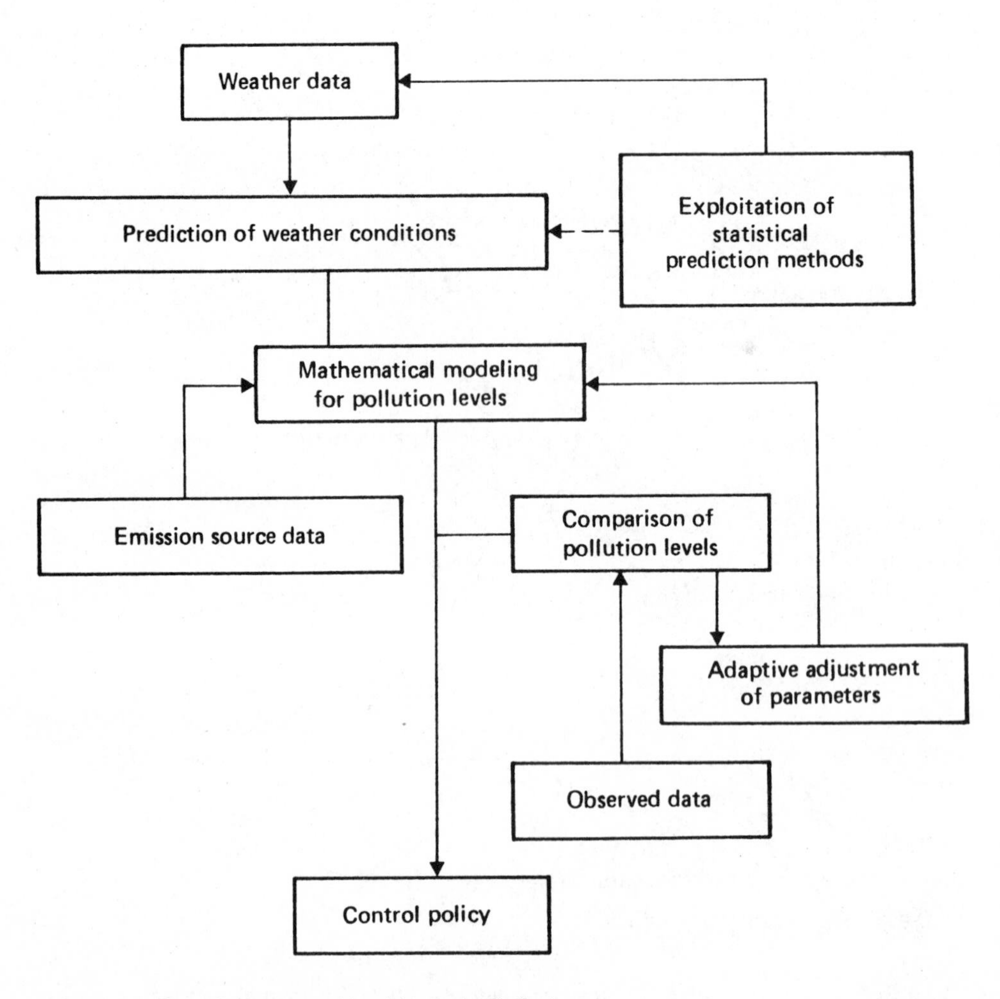

FIGURE 7 Control scheme considered by Osaka Prefecture.

The most important cause of pollution in Osaka Prefecture is the burning of heavy oil, and therefore the concentrations of sulfur oxides are higher than those of other pollutants. Thus, in the above equation "concentration" refers to the concentration of sulphur oxides only. $\bar{p}$, $\bar{q}$, and γ are diffusion parameters, depending upon wind and other weather factors. Although a conspicuous number of studies on the relationships between diffusion and weather conditions have been published, no definite conclusions have yet been drawn. Hence the diffusion parameter γ has been assigned in accordance with the criterion of minimizing the mean-square forecast error.

After model building is complete, a control policy must be assigned, and in particular an objective function for control must be set. The goal suggested by Osaka Prefecture is to maximize the "benefit to both enterprises and inhabitants"; stated in another way, the goal is to minimize the "sum"

loss of polluters due to emission control + loss of inhabitants due to pollution

The first term is the economic loss due to replacing the heavy oil by a cleaner one or to reducing production levels. It is very difficult to measure the second term; therefore the Osaka Prefecture is presently oriented to consider only the first term, while the second term is basically dealt with by setting maximum permissible pollutant concentrations.

3.3 Tokyo Case Study

Air pollution is also a serious problem in Tokyo. For this reason, Tokyo Metropolitan Area devised a measurement system in 1964 and since then has recorded pollution levels and meteorological variables at Tokyo Tower and at Kawaguchi broadcasting center. However, as yet, only data analysis has been carried out, but no control has been enforced.

In 1968 a research group was created and charged with responsibility for data handling and statistical processing. By means of such analysis, mathematical relationships between emission sources, meteorological factors, and pollution phenomena have been established. Thus, the modeling approach in Tokyo has differed from that in Osaka. The Tokyo approach consists of black-box modeling, based on time-series analysis techniques and not "deterministic" modeling, based on diffusion equations. The research has aimed at comparing different statistical modeling approaches and carrying out factor analysis. The results have indicated relationships between levels of several kinds of pollutants, pollution spatial patterns, and weather factor patterns. However, due to lack of data, the analysis has not yet been completed and the results are expected to improve as data records increase.

As regards control measures, in 1970 Tokyo Metropolitan Area decided to set up an air pollution control center, in order to centralize all data measurements and make processing easier, and to establish whether polluters acted upon warnings.

5 AN URBAN CASE STUDY: HIMEJI CITY

5.1 Outline of the Pollution Problem

Himeji City is located in the center of Harima Plain which is in the western part of Hyogo Prefecture. The city has a population of ca. 420,000 and covers an area of ca. 268 km^2. The northern part of the city is a hilly area, while the southern part of the city faces the Seto Inland Sea. The eastern and western parts are mostly hilly.

Sulfur oxides are the main pollutants released into the Himeji air. Such chemical compounds are produced by combustion of the sulfur contained in fuels and raw materials, in power plants, refineries, steel mills, and other plants. Each of the 504 polluting sources is registered at the Hyogo Prefectural Office or at the Himeji Municipal Office and must operate in compliance with the National Air Pollution Control Law as well as within the prefectural air-pollution control regulations. Thus, detailed data on sulfur oxide emissions (temperatures of released gases, emission rates of exhaust gas, and gas emission speeds) are available.

Power plants, gas works, steel mills, and other plants account for about 90% of the total sulfur oxide emissions in the city. Hence Hyogo Prefecture Office has paid most attention to setting regulations governing the building and the expansion of such industrial emission sources, and encouraging the use of low sulfur-content or alternative fuels. At the same time, Hyogo Prefectural Office has strengthened its administrative control and guidance, obtaining the cooperation of many corporations and plant owners. As a result, significant reductions in SO_x levels are being achieved.

5.2 Computer Simulation Model for Himeji City

The reasons why Himeji City was selected for testing the model described herein are as follows. Firstly, from the viewpoint of air pollution, the city is almost completely isolated from other areas. There are no significant industrial concentrations in the adjacent area, except Kakogawa City and Takasago City, which contribute little to air pollution in Himeji City because of their distance. Secondly, most of the city, except for its peripheral areas, is on a plain. The urban area is a gentle slope descending gradually from the north down to the south. This circumstance makes Gaussian modeling appropriate and straightforward. Thirdly, it is easy to obtain data on emissions sources. As mentioned above, Himeji Municipal Office has a large stock of this kind of data, as well as other data collected for administrative guidance. Furthermore, a great deal of meteorological data can be obtained from Himeji Meteorological Station. From mid-1973 on, data on the vertical distribution of air temperatures also became available. Finally, it is possible to obtain additional data since many sensors have been installed in the city. By June 1974, there were 30 instruments for the automatic measurement of sulfurous acid gas, including nine units installed by the city (eight send data to the control center by means of a telemeter). In other words, one unit is operational for every 8.9 km^2. Furthermore, the number of sensors available has since increased.

5.3 Conversational Computer Simulation System (CCSS)

The Gaussian plume model mentioned in the block diagram of Figure 8 is as follows:

$$C(x,y,z) = \frac{Q}{2\pi\sigma_y\sigma_z U} \exp\left[-\frac{y^2}{2\sigma_y^2}\right] \exp\left\{\left[-\frac{(z-H_e)^2}{2\sigma_x^2}\right] + \exp\left[-\frac{(z+H_e)^2}{2\sigma_z^2}\right]\right\}$$

where U is wind speed, Q is emission rate, H_e is effective stack height, and σ_y and σ_z are diffusion parameters. Surface concentration ($z = 0$) is evaluated by the equation:

$$C(x,y,0) = \frac{Q}{\pi\sigma_y\sigma_z U} \exp\left[-\left(\frac{y^2}{\sigma_y^2} + \frac{H_e^2}{\sigma_z^2}\right)\Big/2\right]$$

The model has been tested several times, with satisfactory results.

The data used for the conversational simulation scheme of Figure 8 are as follows: area data – including scales of maps used and characteristics of particular areas in the maps; meteorological data – including details of wind direction, wind speed, and atmospheric stability; and smoke-source data – including data submitted from plants in compliance with laws and regulations, comprising emission rates of sulfur oxides, stack heights, effective stack heights, and locations of the sources.

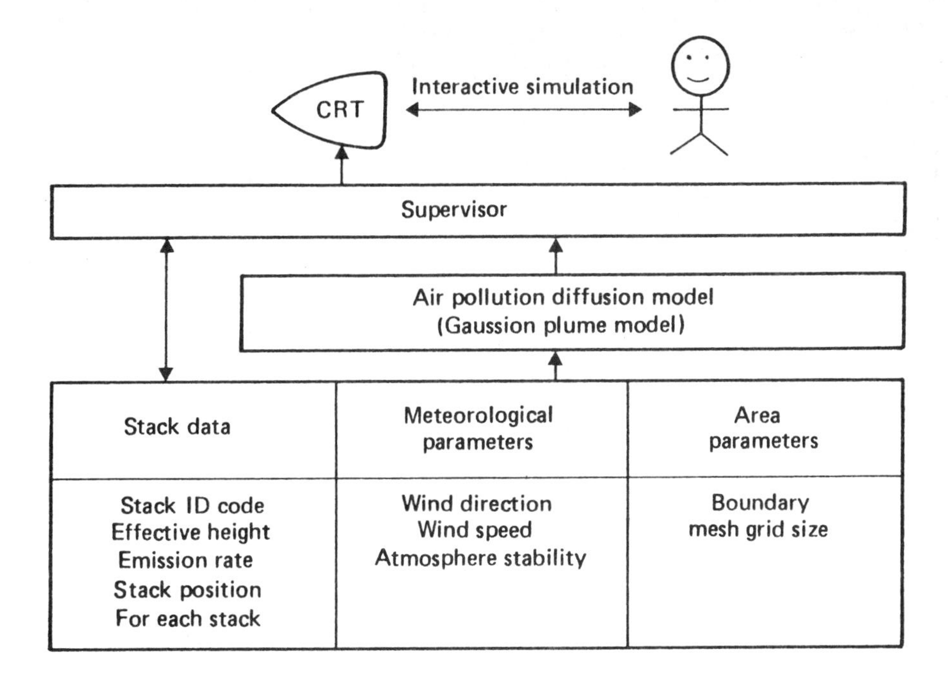

FIGURE 8 Conversational Computer Simulation System (CCSS).

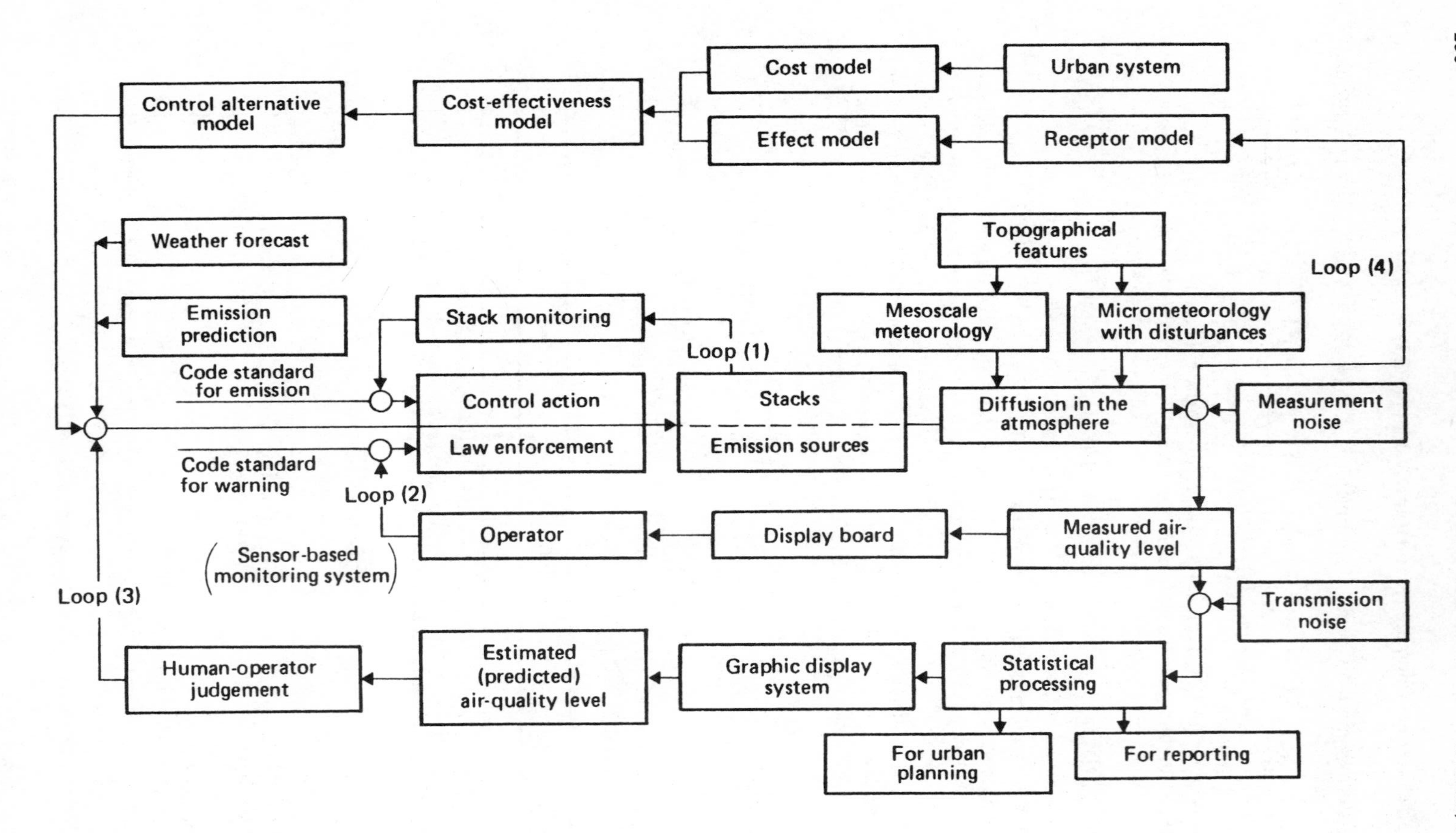

FIGURE 9 Proposed control scheme for Himeji City.

A computer program developed by IBM Japan can display the pattern of air pollution varying on line with changes in the meteorological conditions (the meteorological data can be introduced directly). The general IBM systems approach to the environmental control problem in Himeji City is shown in Figure 9.

5.4 Uses of CCSS

The CCSS can be used for predicting the diffusion of sulfur oxides in three types of situation: when a new smoke source is built or an existing one is expanded; when wind direction, wind speed, and atmospheric stability change; and when the location of a stack is changed. Other prospective applications of the model at the administrative level include the following. The pollutant concentration pattern, as it varies from day to day depending on the meteorological situation, can be computed: the model makes it possible to develop a reliable map of concentration distributions. The effect of prospective smoke sources can be estimated: since the model is multisource, it is possible to estimate the effect of change in the emission rate of each smoke source. In particular, it is possible to predict the concentration patterns resulting from additional (new) emissions. Thus the method can be an effective aid to industrial and housing planning. Finally, the model can be used for environmental control, as outlined in Figure 9.

REFERENCES

APMS (1973). Analysis of Effective Stack Height. Society for the Promotion of Machine Industry.

APMS (1975). Preliminary Report of the Air Pollution Monitoring System. Society for the Promotion of Machine Industry.

Hino, M. (1974). Prediction of Atmospheric Pollution by Kalman Filter. Rep. Jpn. Civil Eng. Soc., No. 224, pp. 79–90.

Yokoyama, O. (1973). The Development of an Air Pollution Monitoring System. Proc. Int. Clean Air Congress, 3rd. Düsseldorf, FRG.

REAL-TIME CONTROL OF AIR POLLUTION: THE CASE OF MILAN

R. Gualdi and S. Tebaldi
Provincia di Milano, Ufficio Igiene e Profilassi, Milan (Italy)

1 INTRODUCTION

Densely populated areas are often situated near industrial zones, so that the superposition of emissions from house-heating and manufacturing plants can cause intolerable pollutant concentrations. Traffic generally represents another source of pollution, particularly when it is slow and runs in narrow streets (which produce a canyon effect). The greatest pollution hazards often occur during the winter months when dispersion conditions are less favorable.

All these characteristics are found in the case of pollution in Milan which is one of the most significant in Italy. In particular, the occurrence of severe and persistent episodes in the Milan metropolitan area during the winter months has made it necessary to have detailed and continuous information about concentration levels.

SO_2 was first selected as a tracer because of the significant contribution of heating plants to air pollution. After a preliminary analysis based on intermittent measurement campaigns, an automatic monitoring network has now been set up: data from sensors are continuously collected and stored in a process computer. Thus the health authorities have a continuous picture of the situation in the whole area.

Because of training and expenditure considerations the development of the network (which is still in progress) has been gradual. Specifically, it has passed through the following stages: (1) monitoring of the urban center of Milan (SO_2), (2) monitoring of the Milan metropolitan area (SO_2), (3) monitoring of the entire Milan province (SO_2), and (4) monitoring of more than one pollutant (NO_x, suspended particulates, etc., in addition to SO_2).

The network is also equipped with meteorological sensors and with noise detectors for airport-noise monitoring. The network has been set up to achieve the following goals (Figure 1): (a) detection of critical concentrations for issuing alarms, (b) determination of potential air-pollution levels (i.e. determination of percentiles of occurrence of critical levels), and (c) identification of correlations and trends in air-pollution levels and meteorological factors, in both the short run and the long run.

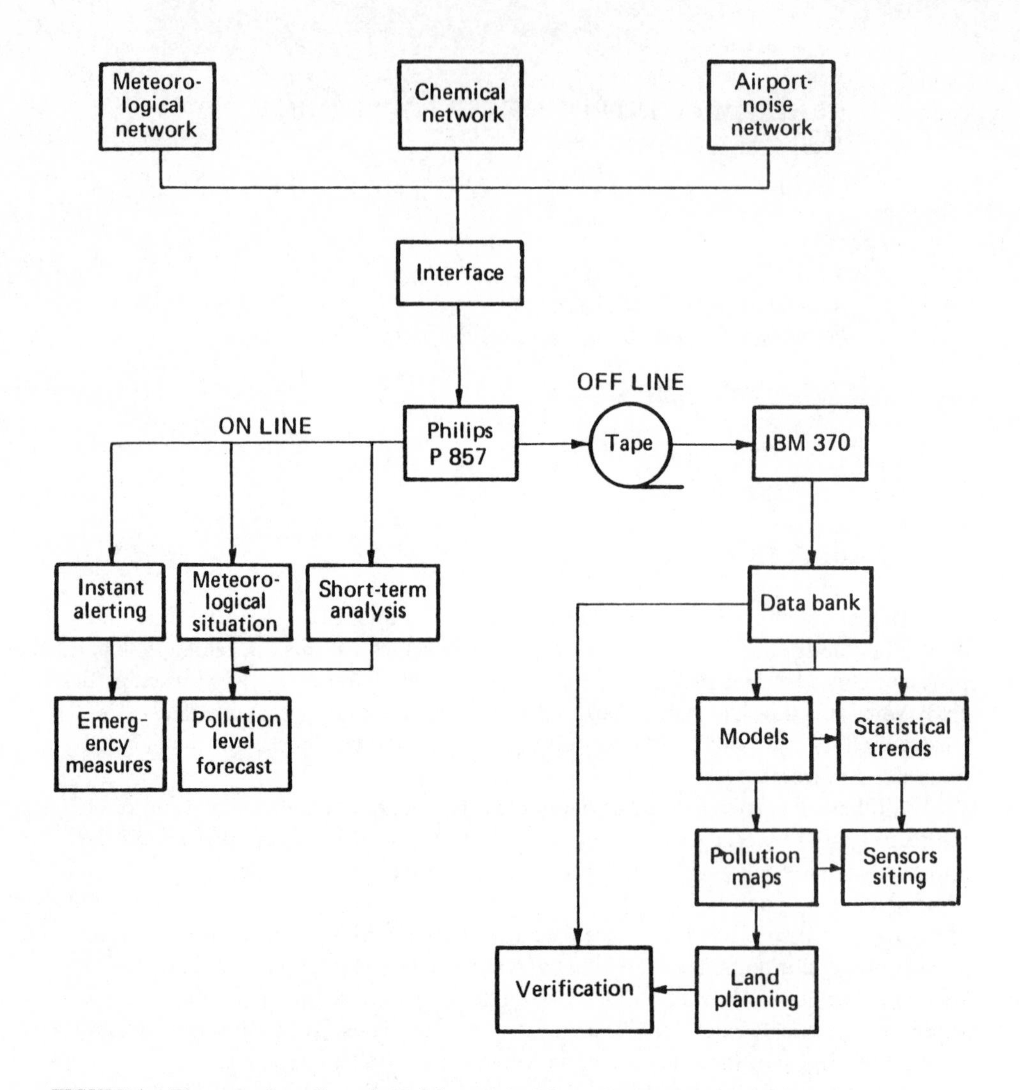

FIGURE 1 The general configuration of the provincial monitoring and control system.

These objectives have in fact been achieved. In particular, the assessment of correlations and trends has illuminated the physical mechanism of pollution in Milan and has allowed the setting up of a mathematical model. After a description of the network (Section 2), the model is illustrated in Section 3.

2 THE MONITORING NETWORK

The network consists of three different types of sensors (the center of the network is located at the Hygiene Office of Milan Province): sensors of chemical-pollutant concentrations, sensors of meteorological data, and sensors of airport noise.

Specifically the equipment consists of the following instruments: 25 SO_2 monitors (ten in Milan and 15 in the province); four automatic dust monitors; six meteorological stations measuring temperature, wind velocity, and wind direction; a tower measuring wind and temperature profiles; five semiautomatic meteorological stations measuring temperature, wind velocity, wind direction, rain, and relative humidity (rain gauge, hair hydrograph); one atmospheric-pressure gauge; one pyranometer; one bimetallic actinograph; one ozone monitor; four nitrogen-oxide monitors; three airport-noise sensors.

In addition the provincial center has one instrument for receiving facsimiles of isobar maps covering the entire European area and one teletypewriter connected to Linate airport for receiving meteorological bulletins.

The sensor locations were chosen initially on the basis of empirical and practical considerations and subsequently by use of a mathematical model. In fact, the first aim in setting up the network was the monitoring of the city center, the most heavily polluted area. Thus ten SO_2 sensors and five meteorological stations were located in this zone, which is approximately a circle of diameter 7 km (Figure 2). The significance of this particular arrangement was then analyzed by a mathematical model; this also suggested both the enlargement of the SO_2-monitoring network to the whole metropolitan area (12 km in diameter) and the monitoring of other pollutants (suspended particulates, NO_x, O_3).

The sensors outside the metropolitan area (Figure 3) are presently located either in highly populated areas or near significant industrial emitters, which the Health Office plans to control in real time (i.e. under unfavorable meteorological or production conditions).

The nucleus of the Milan system hardware is a process-control computer with a 32-K, 16-bit core memory. Standard features of the CPU include 148 micro-instructions, double-length arithmetic mode, 16 hardware registers, 62 interrupt levels, power fail/automatic restart, a real-time clock, and a V24 serial interface control unit.

A general-purpose peripheral connected to the computer includes the following (Figure 4). (a) A display unit which serves as operator's console if required. The operating speed is up to 1200 characters s^{-1}. The screen is 80 characters wide with 24 lines. (b) A cassette tape unit in which each cassette has a capacity of over 6×10^6 bits. The transfer rate is 6000 bits s^{-1}, the recording density is 800 bits in^{-1}, and the tape speed is 7.5 in s^{-1}. This tape unit offers proper work areas to aid software processing. Each item of software is stored on a separate cassette. (c) A magnetic tape unit which operates at 800 bits in^{-1}. The transfer speeds are up to 20,000 characters s^{-1} at a tape speed of 25 in s^{-1}. The capacity of each tape is approximately 40,000,000 characters, depending on the block length and other factors. (d) A line printer unit operating at 200 lines min^{-1}. The print width is 132 characters at a density of 10 characters in^{-1}. The P809 can be connected to a standard interface or to one of the available V24 interfaces.

Finally the computer is connected with a topographic display panel on which each detector (represented by a lamp) indicates whether the average concentration in the last 10 min has exceeded a preset threshold or not.

Any sensor (chemical or meteorological) can be selected and its measured data continuously recorded on four analog records placed in a central control room. By means of the analog display supplied by these recorders it is possible to follow more easily a particular critical pollution episode or to analyze the incorrect working of a sensor.

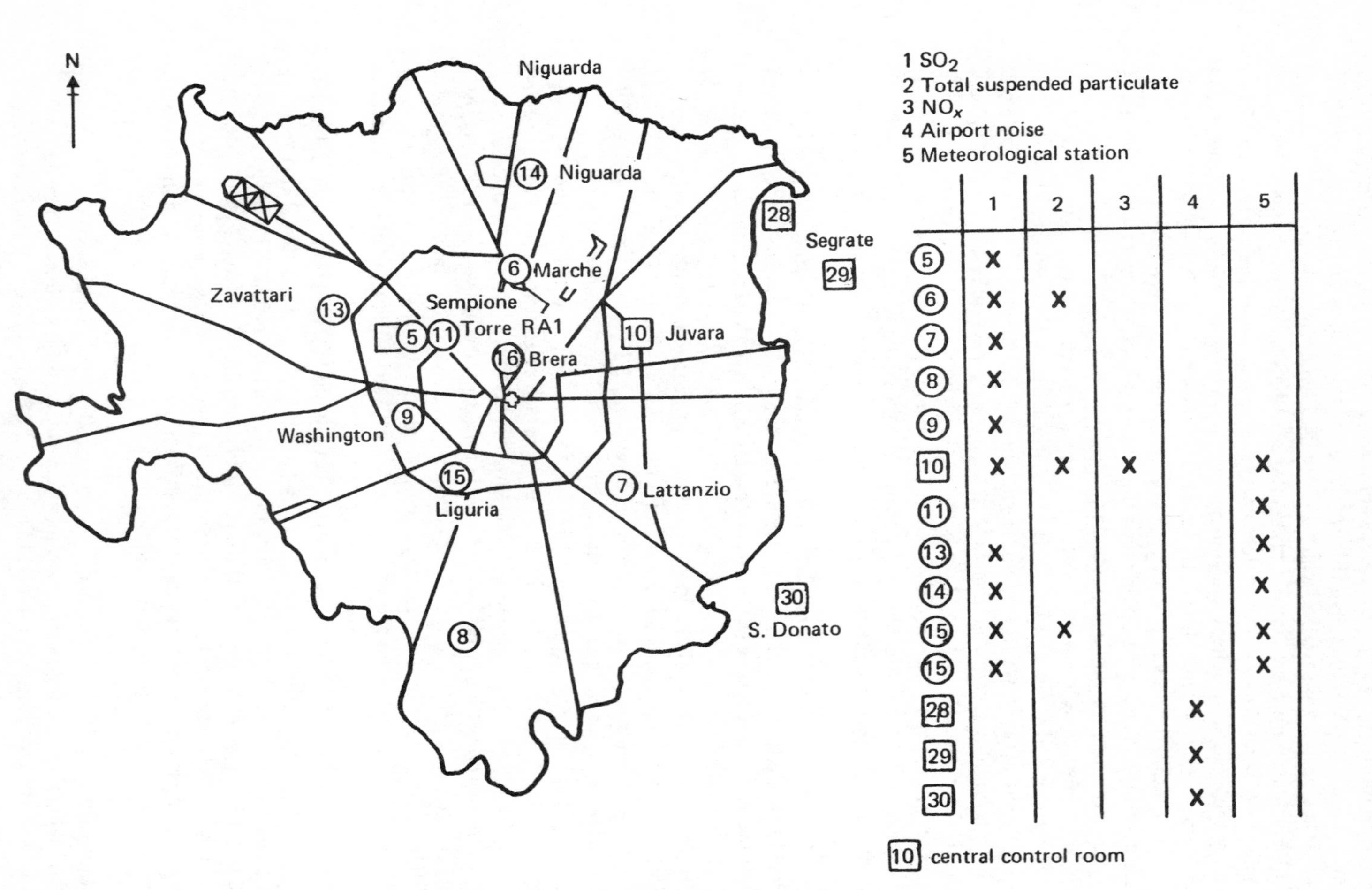

	1	2	3	4	5
(5)	X				
(6)	X	X			
(7)	X				
(8)	X				
(9)	X				
[10]	X	X	X		X
(11)					X
(13)	X				X
(14)	X				X
(15)	X	X			X
(15)	X				X
[28]				X	
[29]				X	
[30]				X	

FIGURE 2 Central and metropolitan Milan showing the sensor locations and the data analyzed.

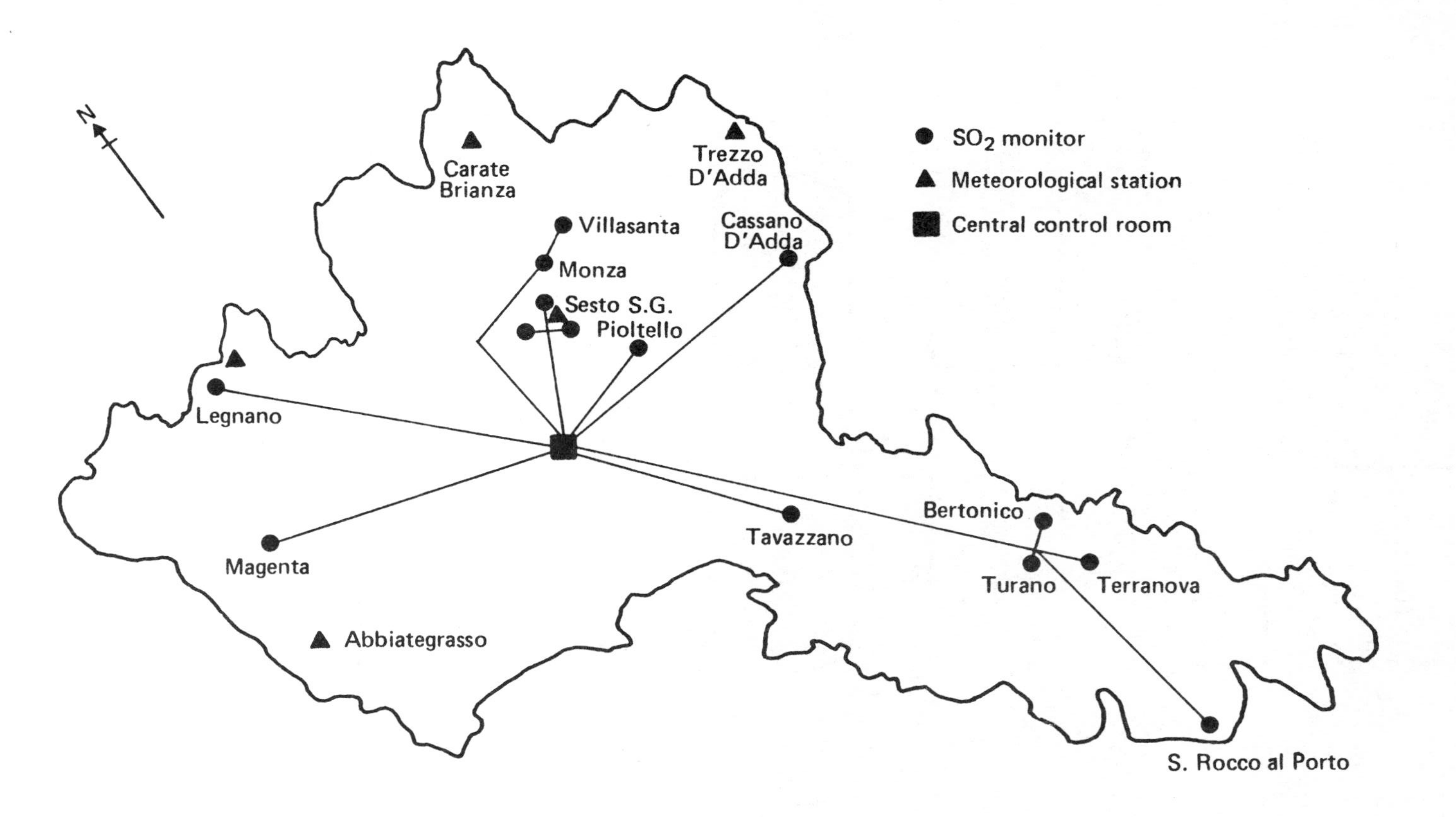

FIGURE 3 Milan Province and the sensor locations.

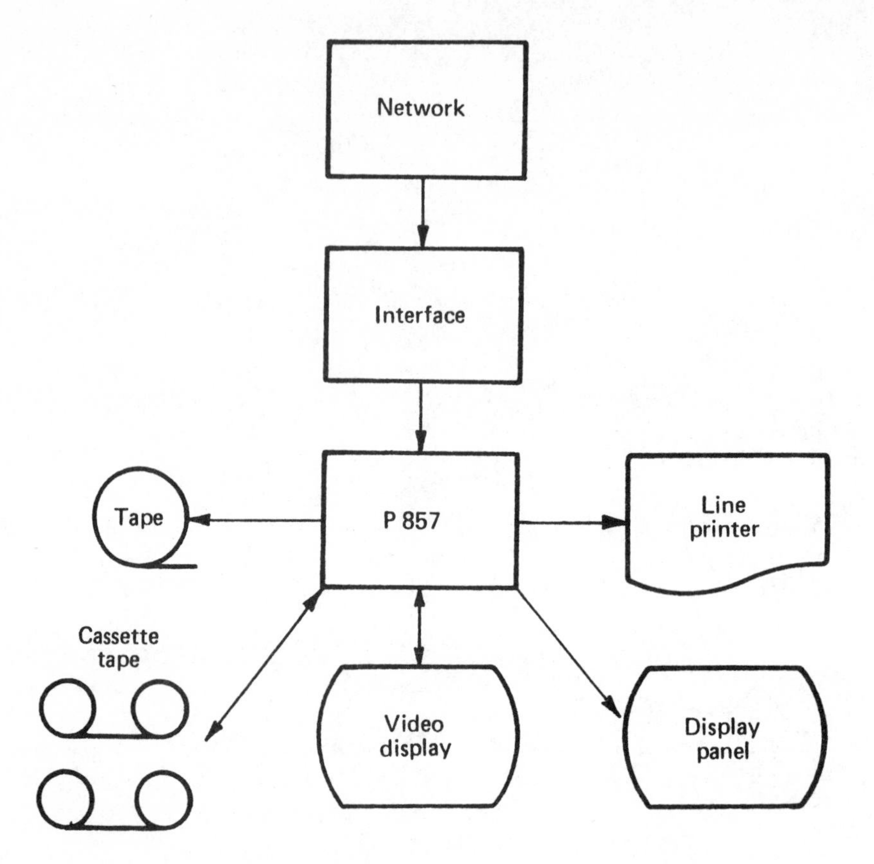

FIGURE 4 The computer subsystem.

The computer is connected via a multiplexer to a MultiTone data transmission (MTT) system. Telephone lines provide the transmission channel between the primary and the secondary side of the data-transmission system.

The MTT system supplies continuous measuring signals (one value every minute) from each of the detectors. After demodulation each signal consists of a series of continuous square-wave voltage pulses.

The pulse frequency of 5–25 Hz is proportional to the actual measured value supplied by the corresponding monitor. The MTT system is based on the frequency-multiplex principle. In this method of operation each signal is allocated to a different carrier frequency (25 maximum) and all the signals can be transmitted simultaneously over one telephone line. A digital signal is transmitted by switching the carrier wave on and off. Analog signal transmission is performed by modulating the carrier wave with a low frequency which is proportional to the analog value.

When a number of analog signals have to be transmitted, one carrier frequency is allocated to each signal. The central computer, guided by built-in programs, supervises the connected equipment (remote monitors, telephone lines, transmission circuits, and interfaces).

When an error occurs the data are rejected; once a day, or more frequently if required, the computer prints a table showing the various kinds of errors that have occurred. The memory of the computer is divided in partitions in which the programs are loaded. The user software for air pollution is aided by a basic real-time monitor which allows supervision of the execution of the programs according to 63 interrupt levels, internal–external interrupts management, management of priority, and insertion of the programs from the timers.

The program of calculation and management of the information is the most important program of the system because it introduces the values from the measurement points connected to the system and transforms them into internal values for the computer.

These values are available directly to the operator or are utilized by other programs for calculations and generation of statistics. This program is composed of a set of subprograms, each executing a precise function and running according to the particular requirement specified by the basic real-time monitor. The single subprograms are for identification and checks of the input signals, calculation and storage of elementary values, calibration request, error management, calculation of averages (30 min, 8 h, daily), airport-noise management, magnetic-tape management, and dialogue with the system.

The operative control of the system is performed by a dialogue program which allows the operator to supervise continuously in real time both the instrumentation and the environmental status. Through properly coded commands it is possible to give information to or receive information from it; to ask for average values, minute values, and calibration cycles; to insert calibration constants; to connect or disconnect some monitors; to dump the central memory; and to ask for the history of errors.

Data reliability and the identification of incorrect data are very important for accurate studies and obtaining real knowledge of the analyzed phenomena. For this purpose, remote detectors particularly suited for automatic networks and able to operate unattended for at least three months have been installed.

In addition an organization has been set up to prevent interruptions to the monitoring and to signal any damage immediately. More precisely, a control system split on the following two levels has been established: (a) automatic actions and automatic checks through the computer; (b) intervention by specialized staff, either for routine preventive work or to correct unexpected damage.

Procedure (a) is a span command (between zero and a preset reference value) to the computer, which by an appropriate program and on the basis of the collected values automatically corrects the measurements (if the parameters of the calibration are within the limits of the characteristics of the instrument). Otherwise the computer points out the malfunction so that it can be corrected. Every day, a 40-min SO_2-detector calibration program is carried out.

Intervention (b) consists of periodic preventive maintenance according to the specifications of the various instruments. This organization gives an efficiency of nearly 90% valid collected data.

The controls of network reliability are printed as tables indicating both hardware errors and incorrect calibrations. At the end of each calibration cycle the computer prints a prospectus containing the measured carrier frequency, the new correction factors, and the error indications (if any).

The measured data received from the detectors are processed on line to compute a general pollution figure. Simultaneously the data are stored on magnetic tape for off-line analysis. If the 30-min averages are greater than the preset threshold the computer immediately signals an alarm situation and prints out all the corresponding chemical and meteorological values. Otherwise the data are stored on auxiliary memories and the printout of the synoptic tables is available once a day (Figure 5).

The philosophy of the system is therefore to signal immediately every alarm condition and to supply all other necessary information. Otherwise, in normal conditions, the computer puts all the collected data onto a magnetic tape.

The available on-line elaborations are calculation and printing of the 30-min average concentrations, printing every 30 min of the meteorological values, command and running of the calibration (zero and end of scale) of all the instruments, automatic warning when an alarm threshold is exceeded (this can be set individually for each station), calculation and printing of the 6–12-h and 24-h averages, analog recording of the concentrations at stations of particular interest, and light display of alarm situations on a synoptic panel.

To facilitate statistical analysis, all the 30-min averages are stored on a magnetic tape (Figure 6). Every value is recorded by a particular code which indicates the correct

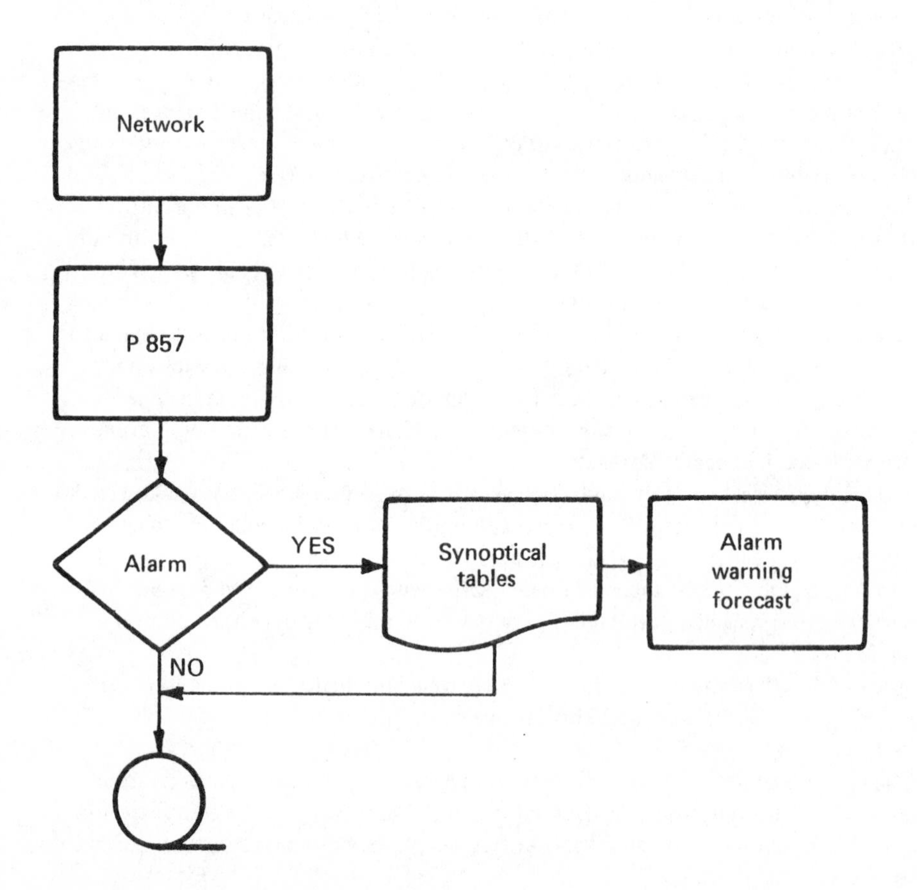

FIGURE 5 The alarm subsystem.

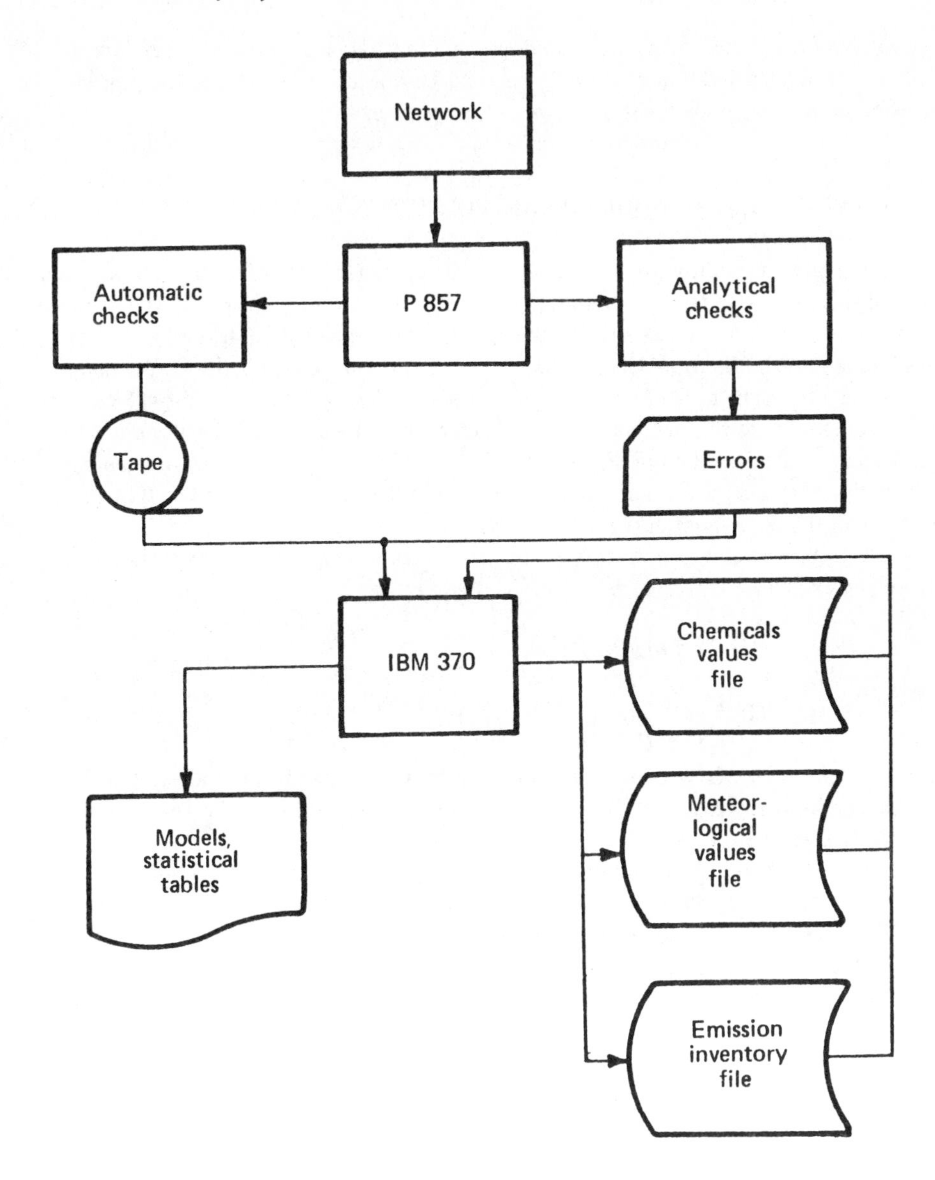

FIGURE 6 The data-bank subsystem.

calibration of the measurement or the correct mode of working of transmission lines. In addition the procedure allows values that are undetected by the automatic check to be eliminated by means of a further analytical control. The purpose of this is to select only available values. Since the computer has a large memory these data are stored on a magnetic disk. This data-management procedure was chosen because of the facilities it offers for data reading, writing, and searching. The file now contains all the chemical and meteorological data recorded since 1971.

Several other networks in Italy are now using this data-management procedure so it will be possible to have an environmental data base available to all the institutions interested in the analysis and control of pollution problems.

3 A MATHEMATICAL MODEL OF AIR POLLUTION IN MILAN

In the previous section we have described the procedure by which the pollution and meteorological file is updated. This file, together with the emission-inventory file from the environmental data set, has allowed the setting up of statistical tables and a mathematical model of the Gifford–Hanna type. The model has been supplied with the meteorological data from the network and with emissions divided into two sets: those from area sources and those from individual sources. The computed values have been compared with the values given by the ten SO_2 sensors. In particular, the whole Milan urban area has been discretized by a grid and each receptor point has been assumed to be located at the center of the corresponding square.

The ground-level concentration x due to an area source has been expressed by Gifford and Hanna as

$$x = (2/\pi)^{1/2} (\Delta x/2)^{l-b} [ua(l-b)]^{-1}$$
$$\times \{Q_A(0,0) + \sum_{i=-4}^{u} \sum_{j=-4}^{u} Q_A(i,j) f(i,j) [(2r+1)^{l-b} - (2r-1)^{l-b}]\}$$

The model calculates SO_2 ground-level concentrations according to the following meteorological conditions: wind speed (m s^{-1}), wind direction (according to the 16 sectors of the compass), and Pasquill stability class (from A to E).

The average wind speed and direction are calculated on the basis of 30-min values given by monitors at the Brera meteorological observatory.

The wind-direction frequency distribution has been normalized in the case of wind calm and the missing data have been proportionally divided between the wind-rose directions. The calculated averages are reliable because local variations of wind speed in Milan are generally not very significant.

The Milan urban area is surrounded by a belt including populated suburbs and industrial installations. Obviously, for an accurate simulation of pollution behavior the contribution of emissions from the outskirts cannot be neglected. Thus the grid pattern also includes the territory north of Milan where conspicuous emission sources are located. The overall area is 400 grid squares of 1 km × 1 km.

In order to supply the model with updated input, emission coefficients (μg m^{-2} s^{-1}) have been ascribed to each square. This procedure is reasonable in the Milan case both because the emissions can be considered as approximately uniform and because of data availability. In fact the available data consist of the volumes of the buildings (constantly requiring updating because of urban development) and their locations. To determine the SO_2 emission the number of installations running on oil and on fuel oil for each square of the network has been considered. From the sulfur content and the fuel consumption of each unit of volume it has been easy to obtain the corresponding emissions of SO_2; by multiplying the unit value by the total volume the emissions of each square are obtained.

This procedure also allows the model to be exploited in land-use planning since ground-level concentrations are directly related to building volume.

For the monthly variation, the comparison between the calculated and observed values is satisfactory (Figures 7–9). To display the surface SO_2-concentration field in Milan, isoconcentration maps have been drawn on the basis of calculated values (Figure 10).

The distribution pattern clearly indicates that there is a main axis of local maximum concentration in the center–eastern side of city corresponding to the highest residential density. Sometimes there is a secondary maximum in the center–western region. Isoconcentrations also indicate the existence of another local maximum in the north of the city, corresponding to highly populated and manufacturing suburbs. In contrast, the low levels in the southeast correspond to a scarcely populated area.

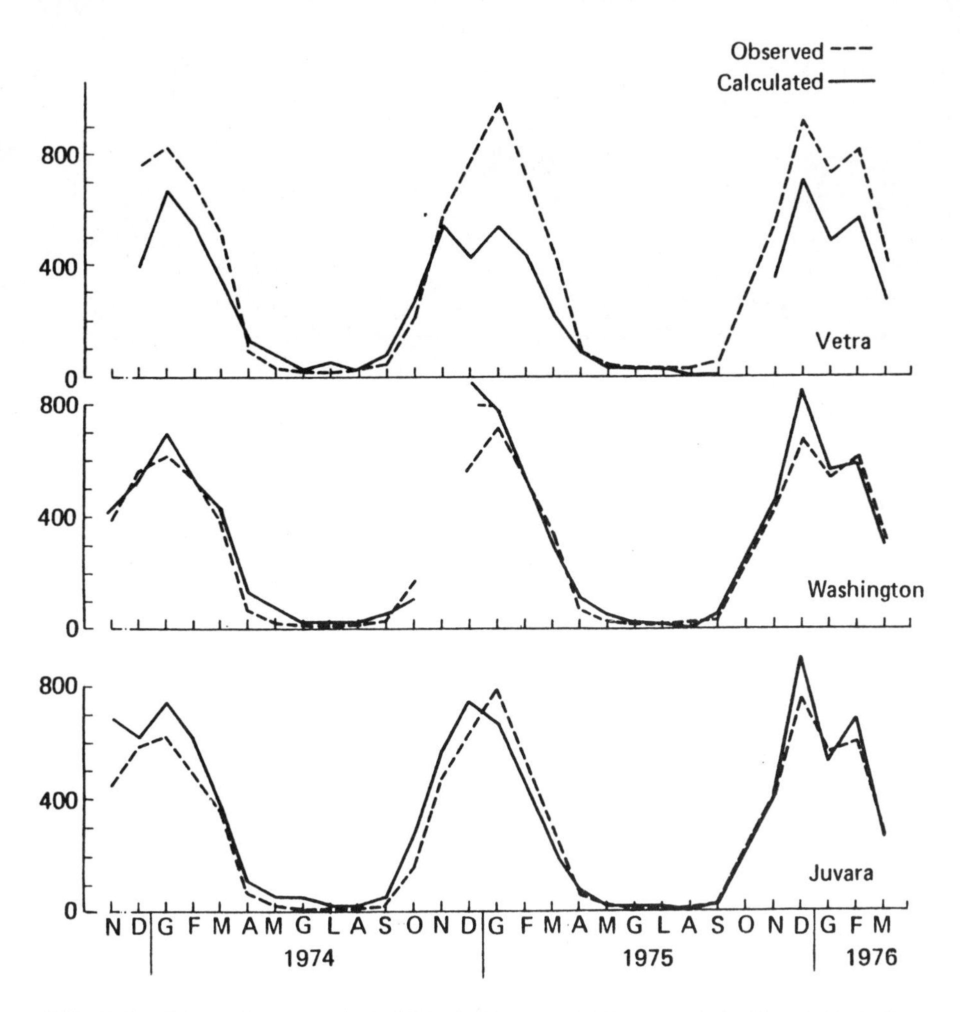

FIGURE 7 The performance of the Gifford–Hanna model in the period 1973–1976 at Vetra, Washington, and Juvara (see Figure 2).

4 CURRENT AND PROSPECTIVE USES OF THE NETWORK

The network is already used for both short-run (episode alarm) and long-run (control of long-term averages) tasks. In addition, on the basis of the concentration levels and meteorological forecasts, a 6-h pollution forecast is sent to the public authorities.

Real-time intervention is not feasible in the presence of distributed sources due to residential heating. However, the information provided by the network and the model actually helps the public authorities to take long-term measures (oil switching) and, in particular, indicates the spatial and temporal extent of the consequences of their decisions.

Moreover, the extension of the network in 1980 to the region outside the metropolitan area and in particular the continuous monitoring of industrial plants will allow real-time emission control of such plants, i.e. the dispatch of fuel-switch orders.

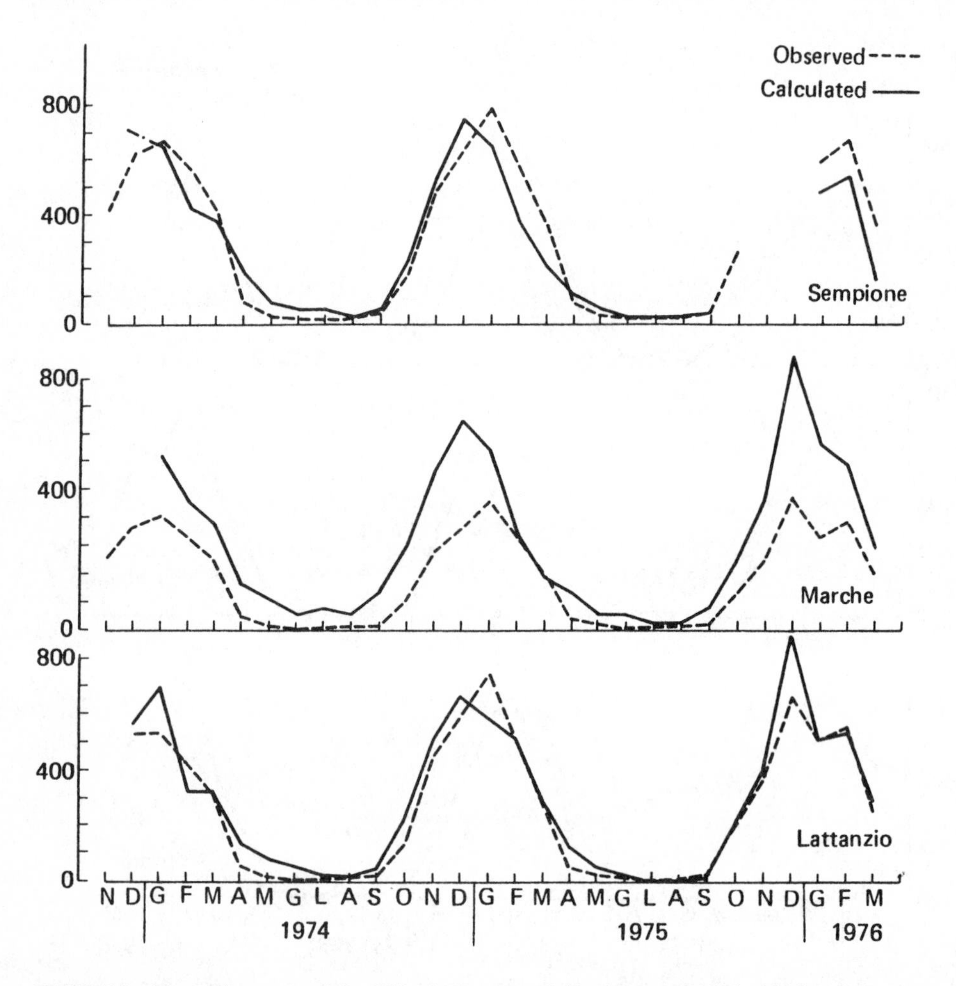

FIGURE 8 The performance of the Gifford–Hanna model in the period 1973–1976 at Sempione, Marche, and Lattanzio (see Figure 2).

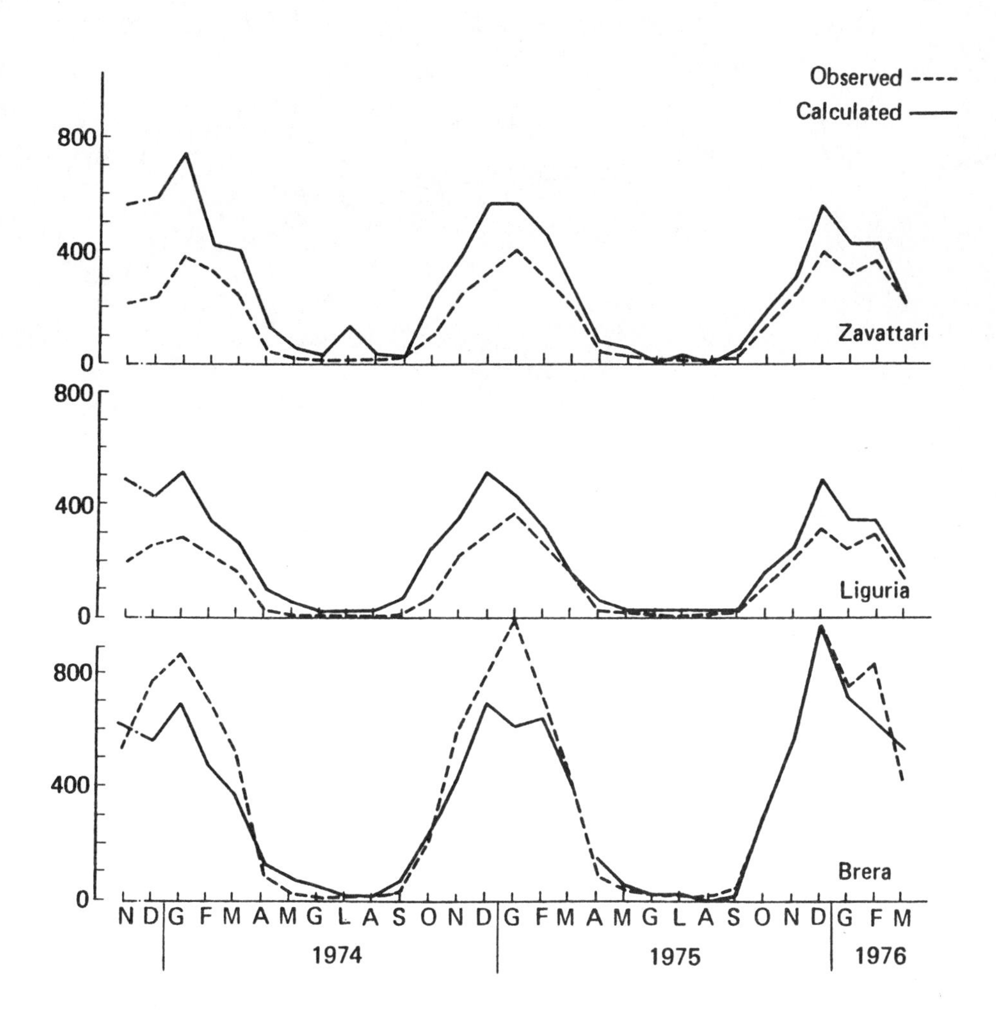

FIGURE 9 The performance of the Gifford–Hanna model in the period 1973–1976 at Zavattari, Liguria, and Brera (see Figure 2).

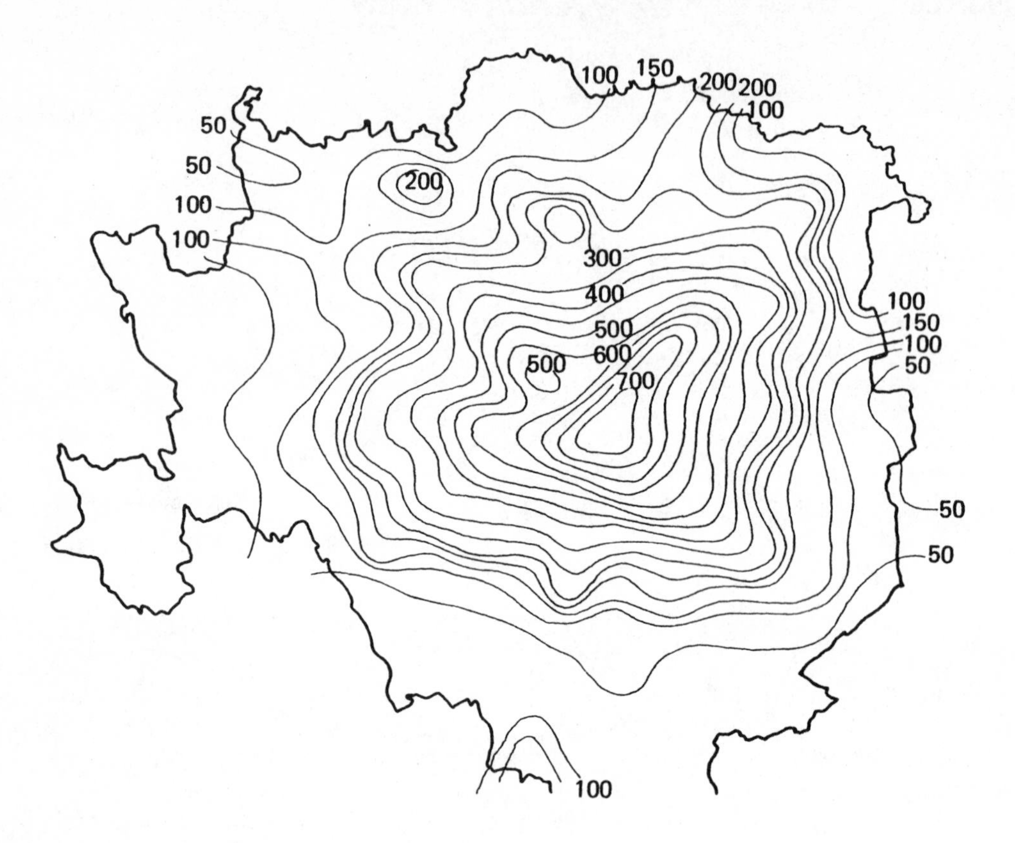

FIGURE 10 An example of an isoconcentration ($\mu g\ m^{-3}$) map.

AUTHOR INDEX